ALSO BY ALLEN ST. JOHN

*The Billion Dollar Game: Behind the Scenes
of the Greatest Day in American Sport—
Super Bowl Sunday*

*Clapton's Guitar: Watching Wayne Henderson
Build the Perfect Instrument*

*The Mad Dog 100: The Greatest Sports Arguments
of All Time* (with Christopher Russo)

ALSO BY AINISSA G. RAMIREZ, PH.D.

*Save Our Science: How to Inspire
a New Generation of Scientists*

NEWTON'S FOOTBALL

NEWTON'S FOOTBALL

THE SCIENCE BEHIND AMERICA'S GAME

ALLEN ST. JOHN AND
AINISSA G. RAMIREZ, PH.D.

BALLANTINE BOOKS
NEW YORK

Published in the United States by Ballantine Books, an imprint of
The Random House Publishing Group, a division of Random House LLC,
a Penguin Random House Company, New York.

BALLANTINE and the HOUSE colophon are registered
trademarks of Random House LLC.

LIBRARY OF CONGRESS CATALOGING-IN-PUBLICATION DATA
St. John, Allen.
Newton's football : the science behind America's game / Allen St. John,
Ainissa G. Ramirez, PH.D.
pages cm
ISBN 978-0-345-54514-5
eBook ISBN 978-0-345-54515-2
1. Football. 2. Physics. 3. Sports sciences I. Title.
GV951.S73 2013
796.332—dc23   2013031521

Title-page image: © iStockphoto.com

Printed in the United States of America on acid-free paper

www.ballantinebooks.com

9 8 7 6 5 4 3 2 1

FIRST EDITION

Book design by Dana Leigh Blanchette

For my grandmother, Gweneth,
who was my life's lead blocker;

and to my mother, Angela,
who has always been
my number one fan
—AGR

For my favorite niece
—ASJ

"Football allows the intellectual part of my brain to evolve, but it allows the emotional part to remain unchanged. It has a liberal cerebellum and a reactionary heart. And this is all I want from everything, all the time, always."

—CHUCK KLOSTERMAN,
EATING THE DINOSAUR

CONTENTS

"Players are ant-like entities," said Stephen Wolfram, straining for an analogy. "Imagine that they have simple rules for interacting with each other—like, if two things are heading straight at each other, they both avoid on the left."

He paused for a moment.

"Even though the rules for the entities are quite simple, the aggregate behavior can be quite complicated."

Wolfram stopped himself.

"I'm sorry, I don't know anything about football," he admitted.

The MacArthur Fellow had been saying this over and over again in slightly different ways for half an hour. But the truth of it snapped into sharp focus when he started comparing Logan Mankins to a bug.

It was a typically atypical day for Team Newton. Throughout our research for this book, we would set up our conference call magic, roll tape on the Olympus digital recorder with the Mickey Mouse–

ear microphones, and talk football with some of the smartest and most accomplished people in the world. One afternoon it was Jerry Rice, who caught a football better than anyone ever. The next it was Lorna Gibson, who can tell you everything you'd ever want to know about why woodpeckers don't get concussions.

It was a great gig.

But the best day of all was when we talked to Wolfram and Sam Wyche.

Wolfram might not be a household name, but in the scientific world he's Tom Brady; his brainpower is matched only by his ambition. His projects range from a cool book that tries to explain literally everything to even cooler software that will do your algebra homework for you. Talking science with Wolfram is as close as you can get to hearing Einstein riff about relativity. But for all his fame, Wolfram is passionate about helping ordinary people use scientific tools to understand the world, which is why he agreed to give us a few minutes on a January evening, eight hours after we hung up with Wyche.

Here's what you need to know about Sam Wyche. The former Cincinnati Bengals coach once stopped a full-fledged stadium riot dead in its tracks with nine words: "You don't live in Cleveland, you live in Cincinnati!" It was the sort of thing that happens in movies, except that this was real life and those were real glass bottles.

That morning last January, we had called Wyche, hoping to talk about Bill Walsh. We were searching for insight on the West Coast offense, the passing attack that revolutionized football in the 1980s. But Walsh himself had passed away in 2007, so we were seeking out his colleagues to fill in the gaps.

Having played backup quarterback for the Bengals and coached the passing game for the 49ers, Wyche had ridden shotgun on one of the most remarkable journeys in football history, a long strange trip from Cincinnati to the Super Bowl. While Walsh tinkered, Wyche watched.

Full of homespun humor and down-home charm, Wyche cordially answered our questions. But he had another story he was itching to tell: his own. So after he finished explaining Walsh's pass progressions in loving detail, Wyche started in on the tale of his own slightly crazy spin on the West Coast offense: the no-huddle.

We were on the verge of interrupting him, because we had another interview scheduled in five minutes. But Wyche is such an engaging storyteller that we simply shut up and listened as he explained why he had decided to toss a century's worth of conventional wisdom out the window. It was a great story, but we didn't know what to do with it. Not yet, anyway.

Our conversation with Wolfram that evening quickly settled into a feedback loop. He would apologize for not knowing anything about football. We would assure him that it didn't matter. He would apologize yet again. We were a little surprised to be offering reassurance to a certified genius, but . . .

Then suddenly Wolfram forgot about football and started talking about chaos theory. It changed everything. "It is a subtle business," he explained, in his lilting English accent. "The knob of the chaos theory idea is the dependence on the initial conditions. Change the initial conditions, and the outcomes diverge exponentially. That is the core of chaos theory."

In this lightbulb moment we each realized that Wolfram was, however improbably, describing what Wyche's innovative game plan had brought to the football field. The no-huddle offense was chaos theory at work.

We understood that *Newton's Football* would be a book not about football or about science but about *ideas*. Ideas coming from the most unexpected sources, converging in the most delightful ways.

On the surface, Wyche and Wolfram are as different as two people can be, the cosmic Odd Couple. During our conversation, for example, each of them talked about how to tell when a football player is getting tired.

Wolfram immediately reached into his bag of high-tech tricks. "It is now possible with image processing to look at your face and see the tiny bits of color every time the blood is being oxygenated as it is being pumped through your arteries. You can measure it with an iPad app," Wolfram explained enthusiastically. "It will measure your heart rate by *looking* at you. There is a whole area of science devoted to determining what a human is thinking from outside. You watch the micro-expressions or, if you could measure it, you look at the skin conductance, or something like this."

Wyche, on the other hand, suggested you look at the guy's thumbs. (It's an old quarterback's trick that we let him explain fully in chapter 9.)

But for all of their differences, Wyche and Wolfram are brothers from another mother. They share an innate curiosity and the intellectual courage to look beyond the conventional wisdom. Both ask "Why not?"

At least when they're not stopping riots or comparing football players to insects.

Why *Newton's* football?

Good question. Set the WABAC machine for 1666. Isaac Newton was out for an after-dinner stroll when he actually did see an apple fall. Through the ages, billions of apples had fallen, but just then Newton was in a particularly contemplative mood, pondering the planets, and he saw in that apple something that no one had ever seen before. He wondered: Why *does* an apple fall perpendicular to the ground? That Falling Apple was a perfect parable for how discovery and innovation happen.

That virtual meeting of the minds between Wyche and Wolfram was a Falling Apple, too. Part serendipity, yes, but it also required a certain openness to possibility.

The history of football is full of Falling Apples.

As is this book.

You'll learn how Otto Graham's mangled cheek inadvertently provided the catalyst for football's concussion epidemic. How a Holocaust survivor's relocation lottery reinvented placekicking. And how an undiagnosed rotator cuff tear gave rise to the West Coast offense.

And of course, there's our own Falling Apple. *Newton's Football* represents a most unlikely collaboration between a writer who, before he wanted to be Bill James, wanted to be Buzz Aldrin, and a scientist who was inspired by watching *3–2–1 Contact* and *MacGyver* and who had a perpetual crush on Mr. Spock. Our other common bond—besides our Hudson County, New Jersey, roots—is a love of explaining complex ideas using down-to-earth examples.

Here's what *Newton's Football* isn't: a comprehensive history of the game or an ordinary book about the physics of the sport. Those have already been done. But it does tell a story, a narrative that begins when the ball was a pig's bladder and continues through today as pro football ponders its uncertain future.

It's a story about innovation. A story about the way small changes can revolutionize the game. And a story about the eternal appeal of the bone-crushing hit, and the corresponding need to make sure that no bones are actually crushed.

Finally, a little hint. Like Wyche and Wolfram, these chapters often bring together elements that at first seem to have little to do with each other. But they do belong together. More than that, they *illuminate* each other. It might take a few pages, but the questions will get answered and the connections made. Just as ours were.

Allen St. John, Upper Montclair, N.J., and
Ainissa G. Ramirez, Ph.D., New Haven, CT.

PART ONE

THE PAST

"Gentlemen, this is a football."

—VINCE LOMBARDI

CHAPTER 1

THE DIVINELY RANDOM BOUNCE
OF THE PROLATE SPHEROID

DeSean Jackson panicked.

The Philadelphia Eagles Pro Bowl kick returner was standing at his own 35-yard line on a chilly December afternoon at newly opened MetLife Stadium in the New Jersey Meadowlands, but the last thing that he expected was the ball.

The plan was simple. New York Giants punter Matt Dodge was going to kick the ball away from Jackson and out of bounds. That might give the Eagles just enough time to throw a Hail Mary into the end zone, allowing the Giants a chance to regroup as this wild game, in which the Eagles had come back from a 24–3 deficit to tie the score at 31, went into overtime.

And then the prolate spheroid did its thing. A bad snap almost fluttered over Dodge's head. Dodge reined the ball in, and his plan was to make contact with the ball just off center and send it toward the sideline, away from Jackson. But Dodge hit it *on* center, just an inch or two away from where he intended. So instead of lofting harmlessly out of bounds, the ball made a beeline for Jackson.

The Eagles ace return man couldn't quite believe what he was seeing. He watched the ball drop, wobbling like a dying quail. The quivering punt was falling to earth just a little short of where Jackson expected it, and instead of taking a step to get under the ball the way his coaches taught him in Pop Warner football, Jackson reached out for it. Starting to run before he had full possession of the ball, DeSean Jackson watched the Wilson "Duke" slip right through the fingers of his dark green gloves.

That's when he began to panic.

With good reason. A herd of Giants defenders were heading toward him at full speed. If the ball bounced forward, one of them would scoop it up, leaving nothing but daylight between the Giants' defender and the Eagles' end zone. If it bounded backward, same deal. But this time the ball fell to Jackson's right. It rolled cleanly, almost gently, end over end the way a small child might, twice, three times, four.

It would probably take the faculty of MIT a full afternoon to parse all the options of where and how that ball might have traveled after it slipped through DeSean Jackson's hands. Almost all of them would have been disastrous for the Eagles.

Instead, the ball rolled around tamely, settling only a yard or two away, where Jackson could pick it up. And thus began the return that came to be known as "The Miracle at the New Meadowlands." Jackson stopped in his tracks, retreating a few steps as he picked up the ball. As he stutter-stepped and pivoted to get going in the right direction, the Giants were in disarray. One defender fell down. Another ran into his own teammate. Still another Giant launched at Jackson and hit another blue jersey.

Jackson looked up, and instead of blue, he saw green—lots of it. The wobbling ball led him right to a crease in the Giants coverage. Now aligned with his blockers, he used his 4.4 speed to motor through the gap and down the right sideline. For dramatic effect, Jackson ran parallel to the goal line before landing in the end zone for the first game-ending punt-return touchdown in NFL history. All because of a random bounce.

It starts with the ball.

Pick up an official NFL football, Wilson Model F1100. It's called the Duke, named after the late New York Giants owner Wellington Mara, who was named after the Duke of Wellington. But forget about that, and forget, too, about Commissioner Roger Goodell's autograph branded into the leather. Instead run your fingers across the pebbled cover. Look very closely, and you'll find a few tiny Wilson logos that are just a bit bigger than the period at the end of this sentence. Explore the stitches that join the four elliptical panels and the bright white laces that play counterpoint to the ball's otherwise sleek silhouette. Pull back for a moment and study its simple, streamlined shape. The ball is nothing if not purposeful. It all but invites you to wrap your hand around it.

But looks can be deceiving. That shape is *anything but* simple. A mathematician would explain that a football is a prolate spheroid. The circumference around its poles (a minimum of 28 inches) is longer than the circumference around its equator (at least 21 inches). The earth, by contrast, slightly flattened at its poles, is an *oblate* spheroid. Why is a football that shape? Ask Arnold the Pig.

Rewind to the mid-1850s. America sits on the brink of war, the game of football is in its infancy, and the very first footballs are made from inflated pig bladders. A pig's bladder itself is relatively small and, after it's removed from the pig, resembles an uninflated balloon. But it's also remarkably flexible, and with proper conditioning, a pig's bladder can stretch to many times its normal size when it's blown up.

As an example of proto-recycling, turning a bladder into a ball is ingenious. But these early footballs were better as symbols than as actual balls. They'd leak or split, and sometimes to keep their shape they'd be stuffed with straw or some other random material. For this reason, these pig-sourced balls fell out of fashion quickly, replaced by balls stitched from leather and rubber—but not before lending a cowhide football its evocative, if incorrect, nickname: pigskin. And not before establishing the unusual shape that would come to define the game of football.

Let's clarify what we mean by *football*. In the days of pig bladder balls, football was something of a pastiche, its rules and style of play very much in a state of flux. The game that would come to be known as "American" football shared common roots with both rugby and the game that Americans call soccer and the rest of the world calls football. We'll discuss the nuances of the game's evolution later, but for the moment, let's focus on the changes in the ball itself. Because in sports, the ball is everything.

In the middle of the nineteenth century, both football and soccer were played with roundish balls that were somewhat irregular in shape because of their origins as, well, part of a pig. As the games diverged, so did the balls. Soccer became a game that centered around kicking. And given the difficulty of controlling the ball with one's feet, one thing became clear: the rounder the ball, the better.

As soon as advances in materials made it practical, soccer moved toward a ball that was as round as possible. The panels became smoother and more uniform, and the seams joining the panels became less prominent. The round balls used in sports like baseball, tennis, and even basketball feature a slight but crucial asymmetry. The orientation of the cover on the round core is what gives Justin Verlander's slider its break and why Kobe Bryant will align a basketball just so—fingers spread across the seams, index finger pointing at the valve—for a free-throw attempt. The cover of a soccer ball, on the other hand, is made up of twelve pentagons and twenty hexagons, the panels forming a figure called a truncated icosahedron, which mathematicians have studied since the days of Archimedes.

Leonardo da Vinci simply called it "divine,"* A soccer ball may not be as purely symmetrical as, say, a billiard ball, but it's *functionally* symmetrical. Players don't care much about exactly where they kick the ball. A modern soccer ball may even be getting *too* round. In pursuit of more perfect sphericity, the soccer balls used in the 2006 World Cup abandoned the traditional thirty-two-panel geometric design in favor of one based on fourteen curved panels. Those sleeker balls were a source of controversy, as the smooth profile allowed the ball to dart in unpredictable ways.† Sometimes, geometry is destiny.

And so it is with the football.

In the latter part of the nineteenth century, rugby—and in turn its cousin American football—diverged from soccer in an important way. It became less about kicking the ball and more about carrying it. And with that, the ball moved away from soccer's trend toward roundness and symmetry in favor of a more elongated shape that honored the ball's porcine roots. The prolate spheroid became even more prolate.

Why? First off, it was easier to carry. A prolate ball could be tucked into the crook of the arm, with a hand positioned over the nose of the ball. Try that with a basketball or a soccer ball and the difficulty of doing so immediately becomes apparent. In rugby, the ball more or less stopped evolving right there. This somewhat elongated ball could be cradled more easily, and even today, the rugby ball remains largely watermelon-shaped.

In football, the evolution continued. In the early part of the twentieth century, the forward pass was first legalized and gradually be-

---

* Three chemists won the Nobel Prize when they discovered this shape in a carbon molecule dubbed buckminsterfullerene for its resemblance to Buckminster Fuller's geodesic dome.
† The eight-paneled Jabulani ball used in the 2010 World Cup featured a textured surface designed to make it perform more predictably, but that ball was also criticized by players for the way it would dart and dive in flight.

came integral to the game. (We'll explore this trend in detail in the chapters to come.) In 1934, the circumference around the ball's belly was reduced from 23 inches to 21 $\frac{5}{16}$ inches and the nose was made more pointed, all with the goal of making it easier to throw.

Wait. What exactly does "easier to throw" mean? For starters, it means being able to throw a ball a long way. The strongest modern quarterbacks can throw a ball 80 yards in the air. But it also means being able to throw the ball accurately; the difference between a touchdown catch and a drive-killing interception can be just a few inches. The modern football addresses both of these requirements, which are sometimes at odds.

Basically there are two forces that a quarterback must contend with as he throws the ball: gravity and air resistance. Here's Newton's First Law of Motion: *An object at rest stays at rest, and an object in motion stays in motion with the same speed and in the same direction unless acted upon by an unbalanced force.* Which means that if a passer could throw a football in a vacuum with no gravity, it would simply continue moving in the direction that the quarterback threw it.

What's an unbalanced force? A good example of a *balanced* force is a tug-of-war between two evenly matched teams. A lot of energy is being expended, but the rope isn't moving. But when you throw a football in the real world, it's all about *unbalanced* forces. The biggest unbalanced force is gravity, which pulls the ball toward the center of the earth—or, in more practical terms, the ground. If you were to graph the flight of a ball, it would describe a smooth parabola, and gravity is the reason. Gravity is a powerful force.

Air resistance? That's a whole different story.

Air is all around us, so we tend to take it for granted, but air resistance is a powerful and largely underrated force. Perhaps the best example of this is the land speed record for a bicycle. Set by Dutch cyclist Fred Rompelberg in 1995, it's a mind-boggling 167.01 mph. How on earth can a bicycle travel that fast? For his record attempt at the Bonneville Salt Flats in Utah, Rompelberg drafted close behind a streamlined car with a specially designed fairing that broke the wind for him. With wind drag eliminated, Rompelberg demonstrated that a human being can indeed pedal fast enough to keep up with a Ferrari in fifth gear. Such is the seldom-acknowledged power of air resistance.

A football also cheats the wind. The elongated shape reduces its frontal area—the surface area exposed in the direction of travel. The smaller the frontal area, the fewer air molecules the ball has to push out of the way as it moves forward, which means less drag. Compare a football with a round ball with the same mass and total surface area and you'll find that the football has less frontal area than the round ball and thus can be thrown farther. Another way of understanding frontal area is by putting your hand out of the window of a moving car. If you face your palm forward, the wind pushes hard against your hand. But if you turn your hand 90 degrees, so that just the side of your hand faces forward, the frontal area is reduced and so is the wind resistance. You can almost feel your hand slicing through the air. This is why sharks and rockets and race cars all have pointy noses.

But if you've ever seen someone try to throw a football for the very first time, you know how hard it is for the ball to maintain this nose-first attitude. Unless, of course, the ball is rotating. When a football spins around its longitudinal axis, this gyroscopic effect stabilizes the ball in flight. This is the same effect that's at work in the world of ballistics, where a bullet can spin as fast as 300,000 revolutions per minute. And it's the opposite of a knuckle ball in baseball, where the ball is delivered with next to no spin. A properly thrown knuckle ball makes only a fraction of a rotation en route from the mound to the plate. That lack of rotation, when the knuckle ball is thrown at just the right velocity, renders the ball susceptible to random air currents, which in turn cause the ball to dart and dive so unpredictably that hitters can't hit it and catchers can't catch it. A football pass that behaves that way is just asking to be picked off.

———

A spiral.

That's the poetic moniker that football fans use for a perfectly thrown, rapidly rotating pass. For all its aesthetic beauty, the real payoff of a spiral is its remarkable accuracy. An ESPN Sport Science segment did a breakdown of the throwing motion of Drew Brees, the All-Pro Saints quarterback, using a ball rigged with sensors. Brees was aiming at an archery target set 20 yards away. His mechanics on the slow-motion replay were not only poetry in motion but also physics in action.

Brees picked up the ball not with one hand, but with two, with his left hand doing most of the work of holding the ball, while the right hand assumed a relaxed grip across the laces of the ball. Brees held the ball in his fingertips, not his palm, leaving a bit of daylight visible in the arch between his thumb and his index finger. Brees's throwing motion began with his weight on his back foot, hips closed, shoulder cocked. As the ball was thrown, his muscles fired and momentum was transferred from the big muscles in his legs and hips, through the rotation of his trunk, into his shoulder, elbow, and wrist in a fluid sequence. It was almost like cracking a whip.

But the magic was still in the hand. As Brees's arm came forward, his wrist snapped. Using the laces for leverage, he put a clockwise spin on the ball. Like Michelangelo's *Creation of Adam,* Brees's index finger was extended heavenward, and it was the last thing that touched the ball. After the ball was gone, his thumb was pointing down, evidence that he had supinated his wrist properly—in other words, rotated it counterclockwise—and his throwing hand arrived safely in the vicinity of his left hip.

Of course, Brees can replicate this smooth and easy motion all day long. Without a defensive end threatening to flatten him, his delivery is so accurate it's almost boring. The instrumented football revealed that Brees throws the ball at a consistent 52 mph, with a 6-degree launch angle and nearly 600 rpm of all-important spin.

Armchair quarterbacks wax poetic about the importance of a

"tight spiral," one where the rotation is perfectly smooth without a hint of wobble. But in this video, the super-slow motion footage reveals that all of Brees's passes have a bit of built-in wobble. When photographed by an ultra-high-speed camera, the nose of Brees's ball actually travels in a tight circle around the ball's horizontal axis, tracing three slight wobbles for every five revolutions. This slight but clearly visible deviation—only a couple of degrees off center—certainly didn't affect Brees's accuracy. From 20 yards out, he was able to hit the 4.8-inch bull's-eye with perfect precision: ten times in ten tries.

If the game's best passer can't throw a perfectly tight spiral, can anyone?

· That's what we asked William Rae. The SUNY Buffalo professor is the world's foremost authority on the flight of the football. He's also the contrarian who debunks the Myth of the Tight Spiral. Watching his own slow-motion footage and studying data from his own instrumented ball, Rae discovered that the flight of the football is more complex than it seems. And that a wobble-free spiral is essentially impossible.

Rae began his research by confirming that a football thrown with little or no spin will do a belly flop once some air gets under the nose. "How do you convince this thing that it ought to continue pointing into the wind?" Rae asks. "The answer is spin."

Or as a physicist calls it, gyroscopic torque. Gyroscopic torque keeps the ball stable in flight, just the way the gyroscopic force of a bicycle wheel keeps your mountain bike upright when you're out for a ride. It also does something more subtle. If you spin a toy gyroscope and lay it on its side and let gravity drop the nose, the gyroscopic torque will move the gyroscope at a right angle, gently resisting the force of gravity.

With a football in full flight, this interaction sets in motion a complex dance between the forces of gravity pulling the ball down and the gyroscopic torque resisting that pull. As the ball spins, there's a cycle of deflection—first slightly to the right, then down a bit, then to the left, then up, and then to the right again. The result is that the spin axis traces out a cone-shaped path called a precession. Which is to say, a spiral with the slightest bit of wobble. Just the way Brees throws it.

Even the gyroscopic torque of a ball spinning sixty times a second isn't strong enough to stabilize the ball indefinitely, Rae explains. If you were to throw a football off the rim of the Grand Canyon, eventually—after ten seconds or so—the cone would get bigger, the wobble more profound, and eventually the ball would start tumbling. In an actual game, that's a moot point because even the longest pass has a hang time of only four seconds, which isn't enough time for the flight of the spiral to collapse completely.

Believe it or not, the inherent asymmetry of the ball has little or nothing to do with this wobble. "The laces have almost no effect," says Rae. He explains that if you were to construct a totally lace-less ball and hand it to a skilled quarterback, and then compare that pass to a throw from the same quarterback gripping a conventional ball on the non-lace side, you'd find no difference between the two passes.

In crunching the numbers determining the trajectory of a thrown football, Rae found something else he couldn't quite account for. A ball thrown by a right-handed passer tends to veer to the right—and by a surprisingly large amount. This deflection can be as much as a yard or two on a 20-yard pass. It took Rae a fair amount of experimentation to realize that this drift wasn't just a glitch in his equations but an actual characteristic of a flying football. But since (a) there are few longitudinal lines on the football field and (b) you rarely see an overhead camera angle of a quarterback passing, almost no one ever notices it.

One who did notice was legendary Green Bay Packers quarterback Bart Starr. A Packers fan who read one of Rae's papers directed the professor to a page in Starr's biography, in which the Hall of Famer explains that his passes naturally trail to the right and that he has to compensate for it.

Shortly after, Rae heard an interview with San Francisco 49ers wide receiver Jerry Rice, just after the right-handed quarterback Joe Montana had been replaced by the left-handed Steve Young. The announcer asked Rice about the differences between the two quarterbacks, and the wide receiver explained that he did detect a difference between their passes but that he couldn't quite put his finger on it.

"It's the gyroscopic torque," the professor could be heard shouting at his television.[*]

The members of Team Madden at EA Sports have a chip on their shoulders. In the gaming world, action-oriented games like Halo and Call of Duty get respect from players and programmers alike. But the guys responsible for the bestselling sports franchise in the video game world are quick to explain that their work is underrated.

"People think that first-person shooter [FPS] games are sexy," says Tim Cowan, EA's group technical director. "But when you look at a lot of other video games, our physics is a lot more complicated."

While dispatching a virtual terrorist with a rocket-powered grenade or blasting a zombie with a 12-gauge shotgun may seem like a taller order than simulating a play in a football game, Cowan and friends will tell you it's not. One of the reasons is that most gamers don't have much firsthand experience with the kinds of weapons brandished in FPS games, so the designers can get away with creating an experience that's entertaining without worrying about whether it conforms to reality. Not so with football. "Every American kid understands what it's supposed to look like when a football bounces," says Cowan. "And when it doesn't look like that, it's very, very obvious."

So you can't blame the Madden programmers for getting a little

---

[*] We interviewed Rice and asked if he made an adjustment for Young's passes. "It is a different spin. If you haven't caught the ball from a lefty it's a big surprise," Rice explained. "I was a little desperate when Steve Young took over. I had a trainer that was a lefty, and I would have this guy throw to me in between practices and stuff like that, to where it became instinctive. I didn't have to think about that spin on the ball anymore."

obsessed with the bounce of the football. They spend hours in offices and hallways and even on the lawns outside dropping, bouncing, and rolling footballs and then poring over the results. They have 10 million lines of code at their disposal, but capturing a football in a way that looks good to the casual observer continues to be a daunting task.

"It's just hard to understand what 'good' looks like," says software engineer Ryan Morse. "If I throw a spherical ball in the air, I know exactly where it's going. If I throw a football in the air and it lands, it can go thirty different ways."

"It's more like thirty thousand. Or thirty billion," adds physicist Toan Pham, the group's technical director.

To see for yourself, pick up a soccer ball, raise it to waist height, and drop it. No surprises there. Drop the ball straight down, and it will bounce straight up toward you. Its bounce is so predictable that after a while you can close your eyes and still catch the ball as it bounces right back into your hands.

Try doing the same thing with a football and watch what happens. The first time, the ball might careen to the right. The second? It wobbles straight ahead. On the third try the ball might squirt to the right again. The fourth, it might veer to the left. Or not. On the fifth, it might bounce backward and hit you in the shin. And those are just broad descriptions. Sometimes the ball bounces up, and other times it wants to squib along the ground. It can come to rest two feet away from where it was first dropped—as it did on Jackson's fumble—or keep rolling until it's ten feet away. Even a few degrees of tilt in one direction or another can send the ball rocketing off in an unexpected way. And a ball that seems to be rolling in a predictable fashion can suddenly change direction on the third or fourth bounce.

"When you roll a football on the ground, it does a lot of unique things," Morse explains. "Sometimes it does exactly the opposite of what you'd expect. You might think it would come to a nice rest and spin about its long axis. But its impetus shifts to the end and it starts to wobble. It always starts to turn. It's a function of its shape."

Play Madden NFL 13 for a while, and it's clear that all of the team's effort paid off.

Sit back and watch as a virtual Tony Romo is sacked by a digital Justin Tuck, and look closely as the Dallas quarterback fumbles the ball. The virtual "Duke" is impressively detailed—you can not only see Roger Goodell's signature, you can also zoom in on the virtual valve stem—but more important, it behaves in a believable way. The ball wobbles as it first hits the ground, and then it squibs away end over end, skidding on the turf, before it spins and finally bounces up so that Romo's teammate Doug Free can recover it.

If you want to quibble, the Madden ball seems to take fewer quirky bounces than a real one, but on balance, the movement of the ball is even more impressively realistic than the action of the players.

What is it that this oddly shaped ball lends to the game of football, both the Madden version you'll find on your Xbox and the one played in stadiums on Sundays? In a word, randomness.

In the real world, randomness is a force that's both ubiquitous and sneakily powerful. Random movements of molecules are at the center of all kinds of chemical change. Random mutations of genes are the driving force behind evolution. Randomness is the engine behind the cryptography that keeps terrorists from sabotaging nuclear power plants and allows you to use your credit card to buy Springsteen tickets online.

In football, the randomness of a bouncing ball adds an element of uncertainty that coaches and players try mightily to minimize. Indeed, the unpredictable bounce of the prolate is powerful enough to determine which teams will be vying for a trip to the Super Bowl and which ones will be watching the playoffs from the comfort of their couches.

Consider the fumble. Forcing a fumble is a skill that can be practiced and learned. And so is covering the ball. But once the ball hits the ground, all bets are off. It will elude the grasp of a future Hall of Famer and leap into the arms of a sub from the taxi squad. The subsequent change of possession—or lack thereof—from a fumble can result in a swing of 14 points or more. In a league where the average margin of victory is 12 points, it's no exaggeration to suggest that one random bounce can determine the outcome of a game or even a season.

In the 2012 regular season, the San Francisco 49ers recovered 23 of 37 fumbles, for a 62 percent success rate. The Detroit Lions recovered only 9 of 24, or 37 percent. Their success at collecting bouncing footballs is one of the reasons why San Francisco went 11–4–1 and made it all the way to the Super Bowl. As for the Lions, they ended up a disappointing 4–12.

In large part because of a certain quirk of a pig's anatomy. And the random bounce of the prolate spheroid.

CHAPTER 2

TEDDY ROOSEVELT IN
THE UNCANNY VALLEY

Football was in trouble. Fans fretted about the number of serious injuries—many of them life-threatening—and the overall level of violence of the game. The pundits protested as well, some of them actually calling for an outright ban of the sport. These problems were serious enough that the president of the United States, a dyed-in-the-wool sports geek, got involved.

While that scenario could be a possible peek into the not-so-distant future—next month's lead story on ESPN, perhaps—it actually happened more than a hundred years ago. The game in question was an early variant of football, played without helmets or pads or even the forward pass, in which players were dying by the dozens from a variety of gruesome injuries. The president who took command of the situation was Theodore Roosevelt. The rules that were hammered out in that meeting formed the basis of modern football. And most important, this meeting marked the beginning of an ongoing give-and-take between the game's powers that be, who make the rules, and the players and coaches, who find ways around them.

Sports fans love creation myths—the first-ever this, the beginning of that. But reality is often messier. Indeed, creation myths fill such a profound psychological need that no one cares very much if they're true. The most famous sports creation myth of all—Abner Doubleday inventing baseball in Cooperstown in 1839—was created out of whole cloth. There's little evidence to suggest that Doubleday, a Civil War officer, cared much about baseball—just a single letter in which he discussed buying baseball gear for the troops under his command—much less invented it. Nor did he spend much, if any, time in Cooperstown. On the date he supposedly invented the game there, he was actually at West Point. But, these facts be damned, the National Baseball Hall of Fame was constructed where Doubleday supposedly invented the game instead of any number of other more plausible places, like the Elysian Fields on the New Jersey side of the Hudson River, where scholars agree the first baseball game was played in 1846. To this day, the game's greatest players are inducted into Cooperstown, not Hoboken.

And so it is with football. Like baseball and virtually every other sport, football wasn't born so much as it evolved. In the early nineteenth century, students on college campuses played a number of ball games that resembled contemporary football. Games like ballown at Princeton, old division football at Dartmouth, and the annual "Bloody Monday" contest at Harvard shared the basic concept of a large number of players trying to move a ball toward a goal. With only a few basic rules, the games got rough and often downright violent. Amid a public outcry, Yale abandoned its game in 1860 and Harvard followed suit the next year. Later in that decade, East Coast schools began playing a variant of these games that incorporated less "carrying" of the ball and more kicking. Indeed, what's generally considered the first college football game, a match between

Rutgers and Princeton on November 6, 1869, in New Brunswick, New Jersey, looked more like soccer than modern football.

Off campus, the Oneida football club began playing a game that incorporated elements of rugby and soccer, and this variant, called "the Boston Game," moved the sport further toward its current incarnation. The game was played on a large rectangular field with a goal at either end, and the object of the game was to advance the ball toward that goal.

In 1873, representatives from Yale, Columbia, Princeton, and Rutgers met in New York to settle on a set of common rules for the game now known as football. Until that point, rules were agreed upon in an ad hoc fashion by the teams participating in any given game. Harvard's team, for example, preferred a more physical game, with rules that were closer to rugby, and the first Harvard-Yale game in 1875 was played under these so-called Concessionary Rules.

A Yale halfback who played in that game, Walter Camp, modified the rules of the game further. Camp could be called football's Abner Doubleday, but his claim to the game's paternity is far more legitimate. He suggested reducing the size of the teams from fifteen to eleven, a change that emphasized skill over brute force. Camp also created specialized positions, many of which are still used today, and codified the scoring system. Instead of starting play with a rugby-style scrum, Camp created a line of scrimmage, and players would arrange themselves in orderly offensive and defensive formations.

In 1882, Camp also devised a system of "downs" similar to modern football. Why? Because, for the first time—but hardly the last—a rule didn't change the game as its author had intended but instead had a radically different effect. Camp's scrimmage rules were designed to open up play and increase scoring. In practice, the exact opposite occurred. In their contest against Yale that year, Princeton realized that they could simply sit on the ball for an entire half—a perfectly legal tactic under Camp's rules—protecting a lead by functionally running out the clock while their opponents looked on helplessly. Camp's subsequent rule eliminated the glaring loophole that Princeton had exposed. Without this innovation, football would have likely died of boredom.

Camp's second attempt at recasting the rules of the game injected another crucial aspect into football: science. Camp's rules essentially divided the game into a series of discrete "plays." This decision, designed as a quick fix for Football 1.0, forever altered the way football was played. Camp's rule changes provided a giant pause button. Soccer and rugby—as well as more modern sports like basketball and hockey—feature a continuous flow of action, with decision-making occurring on the fly. But Camp's version of football was played at a much more deliberate tempo. Football would rely less on instinct and improvisation than on coordinated action. Players and coaches would use the time between downs to draw up plays and discuss their implementation. They would strategize and counter-strategize. They would think. They would *plan*.

Camp's new rules paved the way for a short-lived but innovative strategy, the Flying Wedge. Widely considered the first modern football play, it also combined the two often competing elements that would define the game as it evolved: chaos and order.

"The Flying Wedge was hailed by some as a positive reflection of scientific thinking applied to football," wrote Scott McQuilkin and Ronald Smith in a 1993 article about the play that appeared in the *Journal of Sport History*. "Coaching staffs had been expanded to the extent that the better teams employed specialized position coaches. The strategic progress of college football rested with these advisors in devising new possibilities for scientific plays and arranging them in a well-ordered series as Harvard, Yale, and Princeton were doing."

The Flying Wedge first took wing in the 1892 Harvard-Yale game. Then, as now, this was a serious rivalry, with roots that extended beyond the football field. Teams considered a victory in the annual game a major point of pride. In the first half of that 1892 contest, the teams had played to a scoreless tie, which gave little indication of what was to come. But then Harvard opened the second half with a truly revolutionary play. It began with Harvard kicker Bernie Trafford executing what was the ancestor of the onside kick. Instead of kicking the ball deep, he bumped it gently and then picked it up. His teammates had lined up behind him, with the larger players on one side and the smaller, faster ones on the other. They ran beside Trafford, forming a giant V pointed toward the Yale Bulldogs. Art Brewer, a speedy running back trailing the play, took a handoff from Trafford running at full speed behind this moving wedge.

This mass of humanity took aim at a single member of Yale's unsuspecting defensive line, a poor sap named Alex Wallis. This Flying Wedge bowled over Wallis and enabled Brewer to gain 20 yards. He failed to score—a Yale alum had seen Harvard practicing the play and reported back to Camp, by now the Yale coach, who warned his players to focus on the ball, not Harvard's strange formation— and Yale still won the game 6–0. But with Harvard's dazzling innovation, a new brand of football was born.

The Flying Wedge was devised by Harvard coach Lorin Deland, who was a rather unlikely innovator. A stockbroker by trade, he never played the game and didn't even see his first football game until 1890, when he was in his late thirties. But Deland immediately recognized parallels between the rival teams and two opposing armies in battle. The Flying Wedge, in particular, was inspired by the military thinking of Napoleon Bonaparte.

"One of the chief points brought out by the great French general," Deland wrote later, "was that if he could mass a large proportion of this troops and throw them against a weak point of the enemy, he could defeat that portion and, gaining their rear, create havoc with the rest."

From then on, football plays would be diagrammed with a specific outcome in mind, and each player on the team would be given an individual assignment within that play. Depending on the game situation and the strengths and weaknesses of the opposing defense, the coach or the quarterback would call a specific play. The route to victory lay in these intricate formations, carefully planned and exhaustively practiced before the contest. Football had become a mental game, and a scientific one, too.

It's no coincidence that football got its start on the nation's oldest college campuses. This approach to the game is straight out of the Enlightenment, the eighteenth-century intellectual movement cultivated, in part, at such institutions. Also known as the Age of Reason, the Enlightenment started with philosopher René Descartes and along the way added politicians like Thomas Jefferson, scientists like Isaac Newton, and artists like Wolfgang Amadeus Mozart. While the precepts of the Enlightenment were quite general in nature, taken collectively they represented a startlingly new view of the world. Departing from the religious precepts that dominated earlier philosophy, Enlightenment thinkers attempted to see the world as it was, identify its problems, and seek out answers. They believed that everything could be improved with an added dose of brainpower. Walter Camp, Lorin Deland, and their contemporaries were sons of the Enlightenment, and the game they devised was catalyzed by reason. In football's early days, every play became an attempt to find order in chaos.

It's ironic, then, that the first fruit of this heady brand of football was the Flying Wedge, a play best remembered for its sheer brutality. A contemporary account in *The New York Times* focused on its physical aspects. "What a grand play! A half ton of bone and muscle coming into collision with a man weighing 160 or 170 pounds!" the reporter gushed. "A surgeon is called upon to attend the wounded player and the game continues with renewed brutality." Also notable is that, in contrast to modern football, instead of being on the receiving end of the hardest hits, the offense was dishing them out.

The game's best thinkers, like University of Chicago coach Amos Alonzo Stagg, immediately saw the potential of the play, calling it "the most spectacular single formation ever." Soon it was in use across the nation. The Flying Wedge caught on quickly, with teams no longer using it only on kickoffs but adapting it to plays from scrimmage as well.

For the next season at Harvard, Deland devised sixty of these "momentum" plays, including a variety of fakes and misdirections in which the ball would be lateraled to a player who would pick his way through the pileup.

From a physics point of view, one important distinction separated the Flying Wedge from the plays that came before and after. Players would interlock arms and actually wear special "wedge belts" that featured handles for teammates to grab on to. All of which rendered the Flying Wedge less like eleven men converging on a common goal and more like a single giant blocker attacking the weakest link in the enemy's line.

Using Newtonian physics to quantify the impact of the Flying Wedge illustrates why it was so effective and so potentially dangerous. Assuming that players in the wedge averaged 200 pounds each, a low estimate of the impulse of the collision—the change in momentum after impact—was 2.5 tons, roughly the equivalent of getting hit by an SUV traveling at 25 mph.

For this reason, the Flying Wedge disappeared almost as quickly as it popped up. Some would-be philosophers compared it, not favorably, to the Theban phalanx, a battle formation employed by the armies of Alexander the Great. The more practical-minded simply tallied the body count. Seeing the writing on the wall, Camp suggested a variety of rule changes: increasing the distance for a 1st down from 5 yards to 10, requiring a punt or lateral on the final down, limiting blocking, and restricting players from changing position before the snap. While some of Camp's refinements would have to wait, in 1894 the rules committee did specifically ban plays in which more than three players went in motion prior to the snap.

The Flying Wedge was dead.

But its grisly legacy would live on.

After the Flying Wedge was banned, other "mass" plays continued. And as teams perfected their execution, the level of brutality continued to rise. Football could have changed direction, becoming a game of finesse, but instead it became a sport in which the movement of bodies and the collisions between them were every bit as central to the action as the movement of the ball. The Flying Wedge established a tone for the game that was both coldly calculating and openly brutal. In that regard, the game became more than a little warlike.

Not that this was considered a *bad* thing, at least at first. A few lone voices, like that of Harvard President Charles Eliot, protested the game's violence, describing it as "a fight whose strategies and ethics are those of war" and calling for it to be banned. But President-to-be Theodore Roosevelt, an advocate for his own brand of rough-and-tumble athleticism, came down strongly on the other side, arguing vehemently against neutering the game. "I would a hundred-fold keep the game as it is now, with the brutality, than give it up," he wrote to Walter Camp. What Roosevelt and other defenders of this rough-and-ready style of football didn't see right away was that this kind of play, left unchecked, could threaten the very existence of the game.

The reason this debate became so pitched—and the problems at its core remained so intractable—is the nature of the game itself. Compare football to baseball. Like all sports, baseball is governed by physics, but on the diamond, the forces generally involve bat and ball, one inanimate object hitting another. Football has a different relationship with Newton's laws. The bodies in motion are still subject to the laws of physics, but those bodies are almost invariably *human* bodies. Therein lies both the game's primal appeal and perhaps its fatal flaw.

Why is football so popular? Psychologists argue that the game's enduring attraction can be traced back to a fundamental need it fulfills. The game serves as a substitute for war and other less socially acceptable forms of violence. Like the gladiators of ancient Rome, the denizens of the gridiron do battle, but in a way that's controlled.

There's a reassuring equilibrium between the chaos of the game and the order established by its rules.

When that balance is disrupted, the results can be devastating.

At the turn of the twentieth century, for the first time but not the last, that balance was disrupted. With offenses running mass plays, players were being gravely injured, some fatally. By the year 1906, the annual death toll had risen to eighteen. As the body count mounted, football took an unexpected detour into the Uncanny Valley.

In animating the blockbuster film *Shrek,* the animators at Dream-Works ran into an extraordinary problem. As they worked on the $100 million animated feature, they continued to develop the technology that allowed them to produce ever more realistic animation. They showed these advanced iterations to test audiences, who responded enthusiastically, until one day they showed a version—the best one yet, the animators thought—that everyone hated.

"When they showed it to an audience of children, the children started crying and freaking out because there was something wrong," *Wired* writer Lawrence Weschler told NPR producer Jamie York in 2010.

With the very future of the film in doubt, the DreamWorks animators tried to pinpoint the problem. The audience, it turned out, was primarily disturbed by Princess Fiona, one of the "human" characters in the fantasy film. The filmmakers soon realized they had unwittingly stumbled into the Uncanny Valley.

What is the Uncanny Valley? It's a concept that was put forth in 1970 by Masahiro Mori, a Japanese roboticist who started studying people's reactions to increasingly realistic robots. As reported on NPR, here's what Mori discovered.

A robot that's 50 percent lifelike? That's a good start.

A robot that's 80 percent lifelike? That's even better.

90 percent? Better still.

95 percent? Totally awesome.

96 percent? Get that creepy thing away from me!

As a robot—or an animated image of a human being—gets more and more lifelike, it suddenly reaches a point at which our subconscious minds no longer register the image as a lifelike replica but as a real human who has been horribly disfigured. It's the same kind of reflexive uneasiness you might feel when you see anything from botched plastic surgery to, well, a clown.*

Mori graphed the Uncanny Valley. The X-axis represents how lifelike the robot or animation was and the Y-axis how an audience responded to the artificial life form. The shape of the line gradually rises until it reaches a peak, at which point the approval rating plummets below zero before rising again even higher as the realism approaches 100 percent.

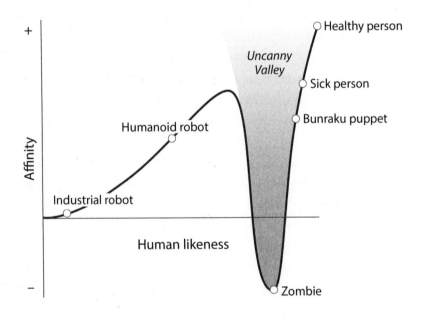

---

* DreamWorks's solution was straightforward: Avoid the Uncanny Valley altogether by making Princess Fiona *less* realistic. Producers of other animated features, like *The Polar Express*, failed to heed this lesson and paid for it at the box office. The animated holiday feature earned only $665,426 in its initial theatrical release, while *Shrek* raked in $267,652,016.

Why? Researchers have a variety of theories about the Uncanny Valley, but the most interesting ones hark back to its evolutionary utility. A human who looks "off" might be ill—and potentially contagious—and therefore may not be a suitable mate. Further bolstering the evolutionary argument, the Uncanny Valley effect is also present in macaque monkeys, who have exhibited the same reaction to too-close-for-comfort pictures of faux monkeys.[*]

What does any of this have to do with football? One of the hallmarks of the game is its intense physicality, often bordering on violence. The game itself has many of the trappings of battle—the object of the game is to encroach on enemy territory, and the strategy consists of attacks and counterattacks. And unlike, say, chess, these moves involve players actually hitting each other.

This violence and our reaction to it can parallel the Uncanny Valley.

When a defensive back tackles an opposing receiver, that draws mild applause from the stands.

A harder hit brings heartier applause.

When the free safety really nails the wide receiver, the stadium erupts.

On the following play, when the wideout gets hit and stays down for a moment, there's a second of stunned silence, followed by even more frenzied applause, some of it for the defensive back and some for the rival player who's walking slowly off the field.

But what about those times when a player gets hit and stays down?

Five seconds go by.

Then ten seconds.

Then thirty seconds.

And the player still doesn't get up.

As the team doctor rushes over to the fallen player, working fran-

---

[*] The adjective *uncanny*, by the way, harks back to psychology pioneer Sigmund Freud, who used it to describe phenomena that are at once familiar and unfamiliar, things we find simultaneously attractive and repulsive.

tically to assess his injury, the stadium goes dead silent. Suddenly the score doesn't matter anymore. That's the lowest point on Mori's graph. The game has stumbled into the Uncanny Valley.

There's an implicit promise made by football to its fans: Players are supposed to get hit. But fans need to believe that they're not getting hurt. A pulled hamstring, a separated shoulder—that's the cost of doing business. A player who's profoundly injured? That's something else. The *controlled* violence suddenly feels *out of control*. The roller coaster has come off the rails. The game is no longer a game.

This kind of reaction isn't restricted to fans. A famous example of this is when Lawrence Taylor broke Joe Theismann's leg on *Monday Night Football*. Taylor, who was one of the most fearsome players in the history of the NFL, never hesitated to do a sack dance over a fallen quarterback. Watch the video carefully, though, and focus not on Theismann but on Taylor. Almost immediately Taylor realizes that this was more than just a routine hard hit. He's jumping up and down, but not in celebration. Having heard the bone snap in Theismann's leg, Taylor is urgently summoning help from the sidelines. In a split second, Joe Theismann went from a being a hated rival with a bull's-eye on his back to a fellow human being writhing on the ground with a compound fracture.

In the last few seasons, this uneasy overlap between seemingly cartoonish violence and real-world suffering has taken on an added dimension with the game's head injury epidemic. Even when the wide receiver gets up after a hard hit, fans are having a more difficult time distancing themselves from reality, as the evidence mounts that at least some of these players will suffer profound and debilitating neurological problems later in life.

"Is it immoral to be a football fan?" asked Will Leitch in an August 2012 story in *New York* magazine. "Can an intelligent, engaged, socially conscious person put the way he sees the world in every other context aside because he enjoys watching the Giants on Sunday? Those are legitimate questions, because you can't just pretend anymore. Every time there's a big hit on the field, I can't keep my human side—the part that wonders what that'll mean for the player when he's forty-five—quiet anymore."

And that's the complicated dance of modern professional football. Hard hits are not only part of the game, they're integral to its appeal. The NFL and its "marketing partners" have long understood this. Television directors give a spectacular tackle as many replays as a circus catch or a touchdown run. Players from Dick Butkus to Ray Lewis have built reputations—and fortunes—on their ability to tackle with no regard for their own physical safety or that of their opponents. And a library of these hits are only a click away on YouTube. The harder the better. Up to a point, anyway.

To his credit, Teddy Roosevelt recognized the magnitude of the problem that football faced, and he took action. Although he never played the game—he was too small as a child and he wore glasses—Roosevelt was a big fan and supporter. He recruited football players for his Rough Riders, and his son played at Harvard. He saw football as a metaphor for the vigorous life he espoused: "In short, in life, as in a football game, the principle to follow is: Hit the line hard; don't foul and don't shirk, but hit the line hard!"

Roosevelt understood that his beloved game had crossed into the Uncanny Valley. It was no longer a rough-and-ready diversion, but a brutal and potentially deadly spectacle that made even rabid fans look away. Legend has it that a picture of the bloodied face of Swarthmore's Robert W. "Tiny" Maxwell spurred the president to action, although no evidence of the photo remains.

On October 9, 1905, he called together representatives from Harvard, Princeton, and Yale, including Walter Camp, as well as administration officials like Secretary of State Elihu Root, to discuss the crisis facing the game.

"Football is on trial," Roosevelt told the representatives. "Because I believe in the game, I want to do all I can to save it."

The outcome of the meeting was the formation of the Intercollegiate Athletic Association, a group initially consisting of sixty-two universities, a precursor to today's NCAA. The group's first mandate: to make sweeping changes to stem the game's injury epidemic. A neutral zone at the line of scrimmage was devised, and gang tackling and mass formations were outlawed. On the positive side, they adopted Camp's earlier suggestion of raising the distance for a 1st down from 5 yards to 10. This opened up the game, requiring teams to do more than simply pound the ball up the middle. One new rule was a bit of an afterthought, but in decades to come it would have the biggest impact of all: the legalization of the forward pass.

These changes marked a decisive step in the right direction, but they were hardly an immediate panacea, as some simplified accounts of this period in football history suggest. The brutality continued through the rest of the 1905 season. In a game against New York University later that year, Union College's Harold Moore died of a cerebral hemorrhage after being kicked in the head while trying to make a tackle. In November of 1905, Columbia suspended its football team, and other major universities like Northwestern, Stanford, and Duke soon followed suit. Fatalities dropped—but only from eighteen in 1906 to eleven in 1907. A subsequent spike in fatalities in 1909 required another round of rule changes.

Ultimately, the new rules did what they were designed to do. Spurred on by Roosevelt's actions, these changes not only steered the game clear of the Uncanny Valley, they laid the groundwork for modern football. Still, this crisis was only the beginning of what would prove to be an ongoing tug-of-war between the players and the rule makers. It's a battle that continues to this day.

CHAPTER 3

THE ROBUST AND FRAGILE FACE MASK OF OTTO GRAHAM

Otto Graham had been through the meat grinder. Or at least that's how a sportswriter described the fifteen-stitch gash to the Cleveland quarterback's cheek in a November 1953 game against San Francisco. This wound would change the course of football history.

Graham is one of football's largely forgotten heroes. He quarterbacked the Cleveland Browns for ten years, leading his team to the title game in each of those seasons, winning seven championships. Graham's Browns started play in the All-American Football League, a league that merged with the NFL in 1950. When the AAFC champion Browns played the defending NFL champ Philadelphia Eagles in the opening game of the newly merged league, Graham orchestrated a stunning 35–10 upset victory that predated Joe Namath's Super Bowl III heroics by almost two decades. While Graham is just a name in the record book to modern fans, serious football scholars and analysts argue that Graham belongs on any short list of the great quarterbacks of all time. Peter King of *Sports Illustrated* called

Graham the best ever, making this elegant argument for his supremacy: "He won seven league titles and seven passing titles in a legendary ten-year career." And yet his most enduring contribution to the game may not be something he did but something he wore: a helmet with pro football's very first modern face mask.

The very first helmet, made of leather, was introduced six decades earlier, in the Army-Navy game of 1893. A doctor had warned one of the Navy players, Joseph Mason Reeves, that he risked "instant insanity" or even death if he suffered another kick to the head. The earliest helmets, like the one worn by Reeves, provided minimal protection and could actually be folded up and put in a player's pocket. Lafayette halfback George Barclay, who is generally credited as the inventor of the football helmet, wasn't concerned with his life or his sanity but with ending up with misshapen cauliflower ears. In 1896, he commissioned a saddle maker to build him a sturdier helmet that consisted of three large padded leather straps. These helmets were called "head harnesses," an apt description.

But the helmet trend was slow to catch on. Most players in the first two decades of the new century didn't wear helmets, and it wasn't until 1939 that helmets were made mandatory. After World War I, football helmets became more substantial, resembling the helmets worn by early aviators. (Indeed, Reeves would go on to become an admiral and is credited with pioneering the use of aircraft carriers, while Navy paratroopers used a variation on his early football helmet.) Still made out of leather, these second-generation helmets were more heavily padded, which improved their ability to absorb impact, and some of them incorporated primitive suspension systems that further isolated the head from shock.

After World War II, materials science was applied to the problem, and plastic helmets that followed the basic design of their leather predecessors were introduced. The earliest plastic helmets were brittle and prone to cracking—Fred Naumetz of the Rams broke nine in a single season—and were actually banned for a year by the NFL while the problem was being resolved. Graham wore one of the earliest plastic helmets, but the design—which lacked a face mask—didn't take full advantage of the new materials.

Until Graham's injury, that is. When Browns coach Paul Brown, one of the game's great innovators, saw the mangled face of his star quarterback, he realized that Graham was one unlucky hit away from being lost for the rest of the season. His solution? Have the team's equipment manager fashion a crude face mask out of clear Lucite that would attach directly to the helmet in the style of a modern face mask. It wasn't a particularly elegant remedy, but it served the purpose. The jury-rigged face mask protected Graham's vulnerable cheek from defenders, Cleveland's star quarterback stayed healthy, and the Browns went on to advance to the NFL Championship Game, which they lost by a single point.

Football scholars have lauded Brown as the architect of modern professional football. He hired the first full-time front office, established in-depth scouting for college talent, analyzed film of games and practices, and housed players in a hotel room the night before a home game, among his many other contributions. In the mid-1950s, he even attached a crude radio receiver to the quarterback's helmet, which presaged today's sophisticated on-field communications. But as important as Brown's other innovations were, none of them changed the game as profoundly as that improvised face mask.

The Lucite face mask was, at best, a stopgap measure. Also known as Plexiglas, the first transparent acrylics were fabricated in the 1930s. The advantages? They were easy to mold and transparent, so Graham's vision wasn't obstructed. The disadvantage? These early acrylics were extremely brittle, almost like glass. Graham was lucky he didn't end up with a face full of plastic shards.

Like so much of sports history, the conventional yarn about Otto Graham's face mask is part truth and part myth. Some college players had fashioned crude face masks during the leather helmet era, many of them resembling a modern hockey goalie's mask, but they never caught on, mostly because of the difficulty of securing a rigid mask to a flexible helmet. A few college players also used face masks on plastic helmets before Graham. A notable example was Johnny Bright of Drake University, who was flattened by a vicious and racially motivated hit by an Oklahoma A&M player in 1951. Bright, who lost his chance at becoming the first African American to win the Heisman Trophy because of his injury, returned wearing a face mask. This was two years before Graham's innovation. But the fact that a pro football star like Otto Graham wore a face mask helped make the new gear acceptable. The plastic face mask was soon replaced by metal models, like the single bar BT-5, which was followed by more intricate designs customized for the player's position. Graham's makeshift face mask would have a profound and lasting effect on the game of football. But not in the ways that you might expect.

Why do boxers wear gloves? Casual fans think the answer is not only easy but obvious. The gloves protect the fighter.

But *which* fighter? And from *what*? It's true that boxing gloves do protect the fighter who's getting hit, at least to some degree. But that's only part of the picture. Bare-knuckle fights can be nasty, bloody affairs, however the earliest boxing matches rarely ended with a life-threatening injury. That's because the most vulnerable part of a fighter isn't his head or even his face. It's his hands. Throw one too many punches and you'll break one of the fragile bones in your hand, which will end the fight right away. That's why bare-knuckle fighters would trade body blows but very rarely punch an opponent in the head. And why smart bartenders of old would wrap a towel around their fist before pummeling an unruly drunk.

Boxing gloves, it turns out, are designed to protect not the head but the hands. But that allows fighters to hit each other in the head for fifteen rounds, or until one fighter is knocked unconscious. Or worse.

And so it is with the modern football helmet. Graham's primitive face mask ushered in a new era of equipment, and soon every player was wearing one. But it became apparent that these improved helmets would be used in ways that Paul Brown never foresaw. In much the same way that boxing gloves made the fist a lethal weapon, the face mask allowed defenders to use their newly protected heads to administer bone-jarring tackles. The gladiators of the gridiron now had their most important piece of armor. The face mask changed the game, but it didn't make it safer.

The reason behind this counterintuitive outcome is two-pronged. On a nuts-and-bolts level, a helmet with a face mask allows a player to spear an opponent with his head. In the pre-face-mask era, the cost of tackling with your head was a broken nose or a shattered jaw. The face mask virtually eliminated that risk.

The other side of the equation is psychological. A phenomenon called compensatory behavior explains why safety devices don't always work as their designers predict. The theory suggests that people have a preset tolerance for risk, and that if a safety device lowers the perceived risk, the individual will—perhaps subconsciously—modify his or her own behavior to reach that internal threshold. In other words, when some device makes you feel safe, you're more likely to behave recklessly. The brain outsources the job of protecting the body to the device, impairing good judgment. For example, drivers behind the wheel of large new SUVs with four-wheel drive and air bags tend to drive faster and more aggressively on snowy roads than they would if they were driving a tiny old economy car. A football player with a helmet and a face mask feels well-protected—perhaps even invincible—and will tackle in ways that he wouldn't even consider with a bare head.

"Football would make a very good case study," says Caltech professor John Doyle. "You're going to see an incredibly rich series of tradeoffs wherever you look." Doyle knows about trade-offs. The university directory calls him Professor of Control and Dynamical Systems, Electrical Engineering, and Bioengineering, but his area of research is dynamic feedback. He studies complex systems—from yeast cells to the next iteration of the Internet—to understand how they work and can be made to work better. It's a relatively new field, and one that remains obscure, mostly because every paper on the subject contains dozens of equations that are incomprehensible to all but a handful of Doyle's fellow scientists. But once you move past the math, Doyle's cutting-edge work addresses real-world questions in ways that are both accessible and thought-provoking.

A record-breaking high school quarterback himself, Doyle is quick to point out that football is a system, and a very complex one at that, in which small changes can have unexpected consequences on a variety of levels.

Two key terms in Doyle's work are *robustness* and *fragility*. Robustness is a system's ability to function even in less-than-ideal conditions. Fragility is a system's susceptibility to catastrophic failure. It is the Achilles' heel of a system and it is the opposite of robustness. Every system has both robustness and fragility operating in a yin and yang relationship.

Doyle uses the example of an aircraft guidance system—his students wrote the software that helped design one—to explain those terms. A modern autopilot system, he explains, is extremely robust to a wide variety of factors, from the changing weight of the plane as it uses fuel to the movement of the passengers, from air turbulence to the length and quality of the runway, all the way down to small factors like the wear on the aircraft's tires. The plane's computer can compensate for a huge variety of variables in ways that even the best pilots can't.

"You can always tell if a pilot's landing a plane or if the autopilot's doing it," Doyle explains. "If it's really, really smooth, it's the computer. And if it bounces, it's the pilot."

That's why we're happy to cede to these guidance systems control of everything from exotic fighter jets to airliners filled with coach-class passengers. The computers do the job better than even the most skilled humans can.

Where's the fragility in an aircraft navigation system? "You're extremely fragile to software bugs," says Doyle. "You're extremely fragile to someone getting in there and hacking the code." He notes that if a sophisticated computer virus were to infiltrate a guidance computer, a plane could fail in ways that would frighten even Michael Bay. Instead of just one plane crashing—as it might if a human pilot made a profound error—a virus could conceivably cause a whole fleet of planes to crash into a populated area. "That's a fragility that's enormous," Doyle says. The key, of course, is recognizing the fragility in the system and taking the steps to address it.

"If you manage all this fragility and robustness properly, you get a net benefit," he explains. "You make yourself robust to very common things. Like wind. And you make yourself super fragile to things that are very unlikely, like a Stuxnet virus taking over all of your aircraft."

The modern football helmet, complete with a face mask, dramatically increased the robustness of the game.

What is it robust to?

Keeping players from getting a broken nose.

Keeping players from getting a busted jaw.

Keeping players from getting a skull fracture.

In short, the helmet prevents the sorts of injuries that knock players out of the game. Those injuries not only sideline players, they're so gory that, as with Graham's mangled cheek, spectators look away. With the face mask, football became more robust to the kind of violence that pushes fans into the Uncanny Valley.

But there's another level of robustness. The helmet allows players to hit hard. And if you take a step back, that's a good thing, at least in the short term.

"We want the spectacular hit. The highlight films in rugby are boring compared to the crazy hits you get in football," says Doyle. "We want spectacular, loud, noisy crashes. But you can't have people getting fractures every time they play. And so you put helmets on."

Hard hits make the game more entertaining. Within only a few years of the near-universal adoption of the modern helmet and face mask, the game exploded in popularity. The 1958 Giants-Colts playoff game, which marked the beginning of the game's Golden Age, featured Hall of Famers like Gino Marchetti and Johnny Unitas, who had a few years to adapt their games to the revolutionary equipment. Defenders of the 1950s and 1960s discovered that hitting with their heads was not only possible, it was preferable. "Bite the ball," coaches would scream, and defenders, leading with the point of their helmets, would do just that.

This technique also left both the defender and the ball carrier fragile in ways that wouldn't be fully understood for decades.

Where are the fragilities of post-face-mask football? The new helmets encouraged players to hit each other in ways that made them far more vulnerable to concussion but offered minimal protection against those kinds of brain injuries.

What exactly is a concussion? Robert Cantu, co-director of Boston University's Center for the Study of Traumatic Encephalopathy and one of the world's leading experts on head injuries, describes a concussion as "an alteration in brain function induced by biomechanical forces." Those biomechanical forces include sudden acceleration and then deceleration of the head, which can cause the brain to crash into the inside of the skull or be twisted or strained in such a way that certain symptoms result. Those symptoms may include, but are not limited to, headache, nausea, sensitivity to light and noise, dizziness, amnesia, drowsiness, the inability to concentrate, and fatigue. Some minor concussions resolve within minutes, while in severe cases a post-concussion syndrome can last for years.

In general, the skull does a good job of protecting the brain against the dangers that an early human might have encountered, like a fall onto soft ground or getting hit with a small stick. Of course, the skull—and the brain it's protecting—fares less well against modern dangers like bullets and motorcycle crashes. Or a 270-pound middle linebacker running at full speed and driving the point of his helmet into your chin.

Why was this considered an acceptable trade-off within the system of football? Because few people knew better. This was the 1950s and the word *concussion* was a term that only doctors used. Sports fans and athletes were only vaguely aware that such a thing existed. They were used to watching cartoon characters like Bugs Bunny and Tom and Jerry getting bonked on the head and "seeing stars" with no lasting effects. Players would actually joke about "getting their bell rung," and teams would rush players back onto the field if they seemed marginally coherent. Punch-drunk fighters suffering from chronic degenerative brain injuries from repeated head impacts had their profound symptoms described using colorful phrases like "cuckoo," "goofy," "slug nutty." Or the evocative "cutting paper dolls."

Concussions were also invisible injuries. They didn't show up on an X-ray. Their diagnoses depended on doctors or trainers asking the right questions and players responding honestly—knowing that bowing out of a game with anything less than a compound fracture might make them look like a sissy to their coaches and teammates. The nervous system usually shook off the acute effects of a concussion pretty quickly, unlike a "serious" injury, such as a broken nose. As long as you were focused only on skull fractures and broken noses, and similar acute injuries, the face mask seemed like a real step forward.

Only now, some sixty years after Graham's innovation, is the problem of concussions being acknowledged and addressed. And we're discovering that the game that once seemed so robust is incredibly fragile to the concussion problem, an issue that threatens the game's very future unless significant changes are made.

CHAPTER 4

VINCE LOMBARDI'S BEAUTIFUL MIND

Vince Lombardi is many things to many people. To football fans, he's the winner of the first two Super Bowls and just may be the greatest NFL coach of all time. To cultural historians, he's an icon of an era when coaches wore fedoras instead of headsets. To Green Bay fans, he's a god.

All of that's true, but Vince Lombardi was one more thing: He was a geek.

A half-century after his glory days, the gap-toothed Green Bay coach is remembered largely in terms of his sideline persona, barking orders at his players and dropping the inspirational quotes that would provide America's motivational speakers with a lifetime's worth of material. While there's a certain truth to the characterization of Lombardi as a charismatic martinet, it's also an incomplete picture. Vince Lombardi was a man of science, aligned in thought and deed with Isaac Newton and John Nash. And once you recognize that, it opens up a richer understanding of the man and the game he loved.

The 1960 NFL Championship Game took place on Monday, December 26, the league bumping the game by a day to avoid a conflict with Christmas. Lombardi's Green Bay Packers were set to play the Philadelphia Eagles. With the benefit of hindsight, this game seems like Green Bay's first step toward becoming one of pro football's greatest teams, but Lombardi had not yet cemented his status as a legend. He had taken over in Green Bay only two years before, and while he made the team competitive almost immediately, the dynasty was still a work in progress.

"That was a pivotal game in our relationship with Coach Lombardi," explained guard Jerry Kramer. "He had come in '59, worked our butts off. We weren't sure about him. We went 7–5 that year; everyone's attitude was, 'We ought to win, as hard as we worked, but is this guy really able to take us all the way?'" While the game didn't answer all of Kramer's questions, it did help Lombardi answer his own questions about how football games are won.

The Packers dominated the game's stat sheet. They had outgained the Eagles 401 yards to 296 and collected 22 first downs to the Eagles' 13. But the final scoreboard told a very different story. The Packers fell short, 17–13.

In the locker room after the game, Lombardi told his team that they hadn't been beaten so much as they ran out of time, a quote that would be preserved in amber in the Lombardi Lexicon. He also vowed that they would never lose another championship game, a promise that he would ultimately keep, as the Packers went on to win five titles between 1961 and 1967.

But after the game, Lombardi sat down privately with Packers play-by-play announcer Ray Scott and offered a more unvarnished assessment of the game: *I blame myself.* He focused on two decisions that, in his mind, lost the game for the Packers. In the first quarter Lombardi had decided to go for it on 4th down and 2 on the Eagles'

5-yard line. Fullback Jim Taylor had gone straight up the middle and was stopped for no gain. The Packers had moved the ball all the way down the field and had come away with nothing to show for it: no field goal, no touchdown, nothing.

A carbon copy of that situation arose early in the third quarter. Green Bay had driven the ball to the Eagles' 27-yard line, where they again faced a 4th down and 2. Taylor again took a handoff up the middle with the same result: Instead of getting three points, the Packers again surrendered the ball. The problem, according to Lombardi, wasn't tactical; he wasn't second-guessing the play call, wondering if Starr should have passed instead of handing off to the fullback. In the coach's mind the error was strategic in nature—why take a chance on moving toward a touchdown instead of cashing in on a makeable field goal? Those 6 points he gambled away, he thought, cost Green Bay the NFL Championship.

Lombardi said to Scott, "I learned my lesson today. When you get down there, come out with something. *I* lost the game, not my players. That was my fault."

"Truth is ever to be found in simplicity," said Isaac Newton, "and not in the multiplicity and confusion of things." Born on Christmas Day in 1642, Newton had a view of the world that someone with a job like Lombardi might find reassuring. The scientist who discovered calculus by the time he turned twenty-five saw the universe as an orderly and harmonious place, a clockmaker's creation. His laws of motion applied to everything, and no matter how complicated the situation, all matter and energy could be accounted for and quantified. No exceptions.

That's how Vince Lombardi saw the world, too; things were subject to reason: They could be explained and measured and, ultimately, perfected.

"If all you have are pushes and pulls—like tackles—then in some sense the whole world should be knowable," explains MIT professor David Kaiser, author of *How the Hippies Saved Physics*. "Once you knew the football players' position and velocity—where they were and where they were headed—then all you had to do was turn your crank on Newton's equations and the whole universe was in some sense determined."

That kind of precision was comforting to Lombardi. He liked the sure thing. And that hard lesson in that first championship game loss only cemented that belief. In the same way that Newton was one of the first modern scientific thinkers, Lombardi brought a modern approach to football, not because his ideas were innovative—he borrowed freely from other coaches like Paul Brown and his mentor Earl "Red" Blaik—but because he integrated these elements into a structured framework.

A deep dive into Vince Lombardi's biography reveals a side of the coach that is largely overlooked today. Before coming to the Packers or the New York Giants or serving as an assistant coach at West Point or at Fordham, Lombardi taught at St. Cecilia High School in Englewood, New Jersey. He didn't teach gym or driver's ed. He taught math. He taught science.

"It was his first football job and his first real job after he flunked out of law school. He taught physics and chemistry," explains David Maraniss, author of the bestselling Lombardi biography *When Pride Still Mattered*. "I think to some extent he was learning along with his students. I don't know if he was a math whiz, but he was more prone to the sciences and math. He was definitely much more comfortable with numbers than words." A counterintuitive observation about one of the most quotable—and quoted—Americans of the twentieth century by the author of his definitive biography.

While the time he spent in the classroom now seems like a footnote in Lombardi's life story, at the time it appeared to be a potential direction for the rest of his career. Lombardi quit his very first job as a repo man almost immediately, but he stuck with teaching, and he was pretty good at it. Lombardi taught science at St. Cecilia for eight years, almost as long as he coached the Packers.

"They all thought he was an excellent teacher," says Maraniss of

Lombardi's students. "He had a capacity to make complex things simple, the same techniques he used in football."

Even as Lombardi moved on to become a full-time coach, he never really left the classroom behind. "They call it coaching, but it is teaching," Lombardi explained, using the somewhat stilted syntax he often employed in interviews. "You do not just tell them it is so, but you show them the reasons *why* it is so, and you repeat and repeat until they are convinced, until they *know*."

Lombardi was a deeply religious man—he would keep a set of fluorescent rosary beads on the steering wheel of his car and on at least one occasion actually dressed in a priest's vestments in his office—and during the 1950s and 1960s the religious and scientific worldviews were not necessarily in conflict. "He wasn't teaching evolution," Maraniss says pointedly.

"He was a sort of dichotomous football coach. He wasn't an automaton like Tom Landry. He very much believed in the human element, in the 'team' and 'love' aspects of football, however corny and clichéd they can be. And was pretty successful in making his players buy into that," Maraniss explains. "He was a very emotional guy. But he was able to carry that emotion along with a very methodological, scientific approach to teaching football. The chemistry and physics teaching that he did at St. Cecilia's, he carried it over into his football life, even to the Packers. In the sense of believing there was a precise, correct, *scientific* way to do everything."

Indeed, it was this gridiron version of the scientific method that helped Lombardi get his foothold into the world of football. When he got his big break—an interview for an assistant's job at Army—Red Blaik, the legendary coach, grilled Lombardi incessantly, quizzing him on the precise details of line play and how to teach it.

"They spent two or three hours, with Blaik asking very, very precise, scientific questions," Maraniss explains. Blaik quizzed him about the spacing between the offensive linemen—should it be two feet or three feet? Do you use the crossover step when teaching the right technique for a pulling offensive guard? "They went through all of these incredibly specific acts of teaching," says Maraniss.

Lombardi's sufficiently detailed answers to Blaik's probing questions helped him get the job that would jump-start his football career. Under Blaik's tutelage, Lombardi would study films, analyzing each player on every play and issuing a grade just as he had in chemistry class. "They were among the first ones," Maraniss explains, "who would try to quantify the science of football that way."

Vince Lombardi came to realize that there was a right, even a *perfect,* way to play football, and the team's collective goal is at least to strive for it. Like Red Blaik, and like Isaac Newton, Lombardi believed in perfection. He believed in answers. And he expected his players to find them.

"Your quarterback days are over," the rookie coach told his Heisman Trophy winner. "You are my left halfback. You are my Frank Gifford. You're either going to be my left halfback, or you're not going to make it in pro football."

That's how Vince Lombardi laid the groundwork for the sport's greatest dynasty. He told the greatest athlete in the game to stick a big portion of his game on the shelf and keep it there. "You're not going to have to worry about playing three positions anymore."

In 1959, Vince Lombardi was a forty-six-year-old first-time head coach facing the seemingly hopeless challenge of leading the worst team in football, the 1–10–1 Green Bay Packers, to respectability. And fast. If he couldn't, he'd get fired and spend the rest of his life as someone else's assistant. Or teaching physics.

Faced with these long odds, Lombardi rejected the idea of a radical new approach to the game. Instead, he fell back on his inherent conservatism and the bias toward precise execution that he learned from Blaik. This was obvious in his unorthodox handling of Heisman Trophy winner Paul Hornung. Hornung was the first pick in the first round of the 1957 draft, selected ahead of future Hall of Famers

Jim Brown and Len Dawson. But by the time Lombardi took over the team two years later, the Golden Boy was seen as a bust.

In the mid-1950s, players were still expected to be jacks-of-all-trades, and Hornung emerged from that mold. At Notre Dame he was a scoring machine, playing both halfback and quarterback as well as serving as the team's placekicker. In this way, Hornung can be viewed as the spiritual predecessor to today's super-versatile quarterbacks, players like Cam Newton, Colin Kaepernick, and Robert Griffin III, who are equally comfortable running and passing. But during his first two seasons with Green Bay, Hornung had split his time among three positions with little success.

"I didn't know where I was playing the first couple years. I played halfback, fullback, flanker," Hornung told the *Green Bay Press Gazette*. "I didn't know from week to week where I was playing."

Lombardi's signature decision was to move Hornung into a specific, specialized role where he'd have fewer responsibilities but be required to execute them at a far higher level. Keep in mind that he did this at a time when the Packers didn't have an established quarterback. Bart Starr was already on the roster, but he, too, was young and struggling. Lombardi decided to tackle one problem at a time. He was happy enough to let defenses *think* that Hornung might throw the ball on the halfback option. But he didn't do so very often, throwing only forty-eight passes in seven seasons, for five touchdowns and four interceptions. Lombardi needed a runner, and he found one. Hornung would do more by doing less.

And it worked. Hornung himself had been so frustrated by his lack of a solid role during his first years in Green Bay that the future Hall of Famer almost quit the game. "I had been ready to bail out," Hornung recalled. "And would have if Lombardi hadn't called." He responded immediately to Lombardi's challenge. He scored seven touchdowns in his first year playing for Lombardi, two more than he had scored in the previous two seasons combined. In 1960, Hornung scored a league-leading fifteen touchdowns. In each of his first three years with Lombardi he was selected to the Pro Bowl as the Packers rose from doormat to dynasty.

Lombardi viewed the passing game with a healthy dose of skepticism. Three things can happen when you throw the ball, Lombardi reasoned—a completion, an incompletion, and an interception—and two of them are bad. If Lombardi put his best athlete at tailback, he installed his most heady and reliable one at quarterback. Like Hornung, Bart Starr had yet to establish himself fully, and the coach obliged by defining his role more tightly.

"Lombardi knew that if he told Starr to do something, it would get done," Maraniss wrote in *When Pride Still Mattered*. "If Lombardi fretted that the quarterback position was too important in football, at least he now had a quarterback who was loyal in every way, who would carry out his game plans flawlessly, who opened up his brain and let Lombardi pour his knowledge in."

That game plan consisted of a lot of handoffs—many to Hornung and most of the rest to Taylor—supplemented with high-percentage passes. Sure, Starr would occasionally throw deep on 3rd and short, but that kind of play call was largely to keep the defense honest. Starr's skills dovetailed perfectly with Lombardi's philosophy. By the standards of the day, Starr was a supremely accurate passer who rarely made mistakes. Between 1962 and 1966, Starr lead the NFL three times in highest completion percentage, lowest interception percentage, and overall passer rating. His career interception percentage of 4.4 percent, while not remarkable by today's standards, made him one of the most precise passers of his era; Starr was picked off less frequently than any quarterback who started his career before 1960.

"Every football team eventually arrives at a lead play. A number one play. A bread-and-butter play," said Vince Lombardi, overselling his message just a bit for the cameras.

In a film with the telling title *The Science and Art of Football*,

Lombardi revealed a little more about himself than he probably intended. Wearing a short-sleeved white shirt, a skinny black tie, and horn-rimmed glasses, he held a pointer in front of a blackboard filled with Xs and Os. Although he was the greatest football coach in the world by this time, he gave off the vibe of a physics teacher trying to engage his students on the first day of school.

"It is the play that the team knows they must make go and the play that its opponents know they must stop," Lombardi enthused. "Continued success with it makes a number one play because from that success stems your own team's confidence, and behind that is the basic truth that it expresses the coach as a coach and the players as a team. They feel complete satisfaction when they execute it successfully." Lombardi paused for a moment, shifting from motivator to instructor. "*My* number one play is the Power Sweep."

These reels of grainy and faded color film are a gold mine. They show both sides of Lombardi, the salesman and the scientist. Like a good used car dealer trying to seal the deal on a late-model, low-mileage Buick, he began by touting the many advantages of the Sweep.

"There's nothing spectacular about it. It's just a yard gainer," Lombardi said conspiratorially. "The Green Bay offense is only as successful as the Green Bay Sweep is. One of the advantages is that we feel we can run it against any defense, be it even or odd."

But once he had your attention, Lombardi sweated the details. Indeed, Lombardi once gave an eight-hour seminar on his beloved Sweep, and a young John Madden left the room thinking that, compared to the Packers' coach, he didn't know anything about football.

The Sweep was the on-field manifestation of Lombardi's worldview, a play that's inherently safe, conceptually simple, yet infinitely tweakable. As taught by Lombardi, the Sweep began as a basic running play. The pulling guard leads the play, sealing off the defensive penetration, while the running back turns the corner and heads downfield into the gap that this precise blocking has created. But God is in the details, and on this count, Lombardi was nothing if not devout.

"If we're looking at this play, what we're trying to get is a seal *here*. And a seal *here*," Lombardi explained, leaving thick lines of chalk on the board for emphasis. "And we're trying to run the play in the *alley*," pointing to the gap between his two lines. As he delved deeper into the play, Lombardi obsessed over the minor variations the offense might employ in reaction to adjustments by the defense, a game of speed chess on the line of scrimmage. He described not only the individual blocking assignments but also broke down the footwork needed to execute these alternate blocking patterns. The details were important, and so was the rigor with which Lombardi pursued them.

The Science and Art of Football was Lombardi's manifesto, demonstrating his belief that football games can be won with relatively simple game planning. As long as that game plan was executed to perfection by players who were just a little bit better—physically and mentally—than their opponents.

"Gentlemen," Lombardi told his players early in his reign, "we're going to relentlessly chase perfection. Knowing full well we won't catch it. Because, in the process, we will catch excellence. I'm not remotely interested in being just good."

A motivational speaker could find a lot to like here. But so could a scientist.

While Lombardi was teaching his students at St. Cecilia, another of the great minds of the twentieth century was toiling in a classroom at the other end of the state at Princeton University. Most people know John Nash as the tragic hero of the Academy Award–winning film *A Beautiful Mind,* the genius with a code-breaking mind who also suffered from paranoid schizophrenia. But Nash's actual intellectual contribution was his pioneering work in game theory. And his epoch-shaking ideas in this field can provide an insight into the theoretical underpinnings of Vince Lombardi's success.

Game theory is basically a branch of mathematics and economics that attempts to quantify real-world problems as though they were games. The problems can range from something as simple as a game of tic-tac-toe to something as large as the workings of an entire economy. Game theory works well in explaining actual games and especially ones that are conceptually tidy, if not actually simple. For example, game theory works better at analyzing grandmaster-level chess, than the economics of a grammar school bake sale. For all its complexity, chess follows rules in a way that a bake sale doesn't.

There's a beginning and an end to a chess match, so it's a *finite* game. Chess is a game of skill rather than a game of chance, so it's a *deterministic* game. There's one winner and one loser, so it's a *zero-sum* game. And because a player understands the options and payoffs available to his opponent as well as his own options and payoffs, it's a game of *perfect information.*

Football, at least as it was played in Lombardi's day, is more like the chess match than the bake sale. It's finite, it's zero-sum, it's largely deterministic, and it has most of the characteristics of a game with perfect information. In that way, it can be understood through game theory.

The simplest contest that game theoreticians use is one called matching pennies. Each player turns a penny over, heads or tails, while hiding it from his opponent. They both reveal their pennies at the same time. If they match, player A gets both pennies. If they don't, the pennies go to player B.

"Yes, it's a really boring game," says Stanford economist Matthew Jackson. "You play it with a two-year-old and they get bored with it. But at a conceptual level, football is a lot like matching pennies."

Only this time, choosing heads or tails would be replaced by the option to run or to pass. And the reward would not be pennies but yardage.

Without realizing it, Lombardi enlisted the thinking of John Nash to attempt to win football games. On each play his offense has two choices, to run or to pass, and associated with each choice is the number of yards that can be gained. On the other side, the defense attempts to anticipate the offense's strategy and selects a defense in response. Each combination of choices has a certain number of yards that might result, which can be put into a simple table (called a payoff matrix), where the results of the offense's choices are listed in rows and the defense's options are listed in columns. Imagine the Packers are 1st and 10 at their own 40-yard line, early in a tie game against the Bears. Lombardi's payoff matrix might look something like this:

		BEARS DEFENSE	
		RUN	PASS
PACKERS	RUN	+3	+7
OFFENSE	PASS	+8	0

If football was just like matching pennies, where the payoffs were equal for heads or tails, Lombardi's choice of plays would be clear and simple: a 50/50 mix between run and pass. But in the NFL during the late 1950s and early 1960s, the payoffs for running and passing were different.

If the Packers called a running play, and the Bears also guessed run, they'd advance the ball 3 yards. If they ran the ball when the Bears were playing a pass defense, they'd move 7 yards.

Passing offered a less appealing set of payoffs. Sure, if the Packers passed when the Bears guessed run, they stood to pick up a nice chunk of yardage. But if Chicago guessed right, on average the Packers would make no gain on the play. And if any given pass results in an interception, that play becomes a *big* loser that can actually swing the outcome of the game.

When we analyze these options it's easy to see why Lombardi liked running plays. If the defense guessed wrong, Green Bay would be close to a 1st down. But even if they guessed right, the Packers would still gain about 3 yards a pop and move the chains down the field.

Coach Lombardi relied on a bread-and-butter running play, the Sweep. His opponents knew it was coming, but he didn't care about it because his team executed it so well. So in relying so heavily on the Sweep and other running plays, Lombardi was actually using a sophisticated spin on conventional game theory: the maximin strategy.

This variation places less emphasis on *maximizing* your potential payoffs, focusing instead on *minimizing* the negative impact of the worst-case scenario.* In practice, Lombardi knew that he couldn't run on every down, but he could devise a mix—running about two thirds of the time—that addressed the situation so ideally that he could advertise it in advance and still make it work. The act of selecting a strategy that succeeds even in an environment of perfect information—where one's opponent fully understands your payoffs—is called a Nash Equilibrium. It won John Nash the Nobel Prize.

Lombardi also recognized something crucial early on. Football wasn't matching pennies or even playing chess. The key difference? Execution. In the late 1950s, pro football was just coming into its own. Not every team had adopted the best practices of rivals around the league, so a team like Lombardi's Packers could earn a significant edge by studying game film or holding two-a-day practices. Those small edges paid off on Sunday, kind of like the house odds in a casino game.

Lombardi's goal was not to outwit the other coach, to win with a smarter game plan, or to defy conventional wisdom by going for it on 4th down. Vince Lombardi wanted to be the *anti*-genius. He would readily sacrifice the chance for a 20-yard gain on a passing play if by doing so he could also eliminate the possibility of a 10-yard loss on a sack, or giving up the ball on an interception. The Packers didn't try to fool anyone with Lombardi's signature Sweep. The opponents knew it was coming, but the team executed it so well that they still gained 3 yards.

"If you're the better team, you want to force the game into low variance plays," says Brian Burke of Advanced NFL Stats, a website

---

* The offense's maximin strategy and their Nash Equilibrium strategy are to run 67 percent of the time and pass 33 percent of the time, while the defense would employ a run defense 58 percent of the time and pass defense 42 percent of the time.

specializing in statistical analysis of professional football. "Football's funny that way. If you can assure yourself three and a half yards on every running play, you'll be unstoppable."

Indeed, Vince Lombardi's idea of a perfect football game—and maybe Isaac Newton's too—looks something like this: running the ball for 3 or 4 yards per play all afternoon, grinding time off the clock to minimize the number of plays, and taking advantage of your opponent's impatience. In 1961 and 1962, the Packers did just that, leading the league in rushing yards and the fewest turnovers, and they would rank near the top in those categories throughout the team's reign. Green Bay's defense also specialized in turning desperate passes into interceptions, leading the league in 1962 and 1965.

The scientist in Lombardi understood the game of football in terms of physical forces and bodies in motion. But he also understood it in terms of probabilities. And the way to triumph in that era was by gaining small advantages through better execution and then adopting a conservative game plan that wouldn't fritter them away.[*]

For all its obsession with numbers, game theory shakes down to an understanding of the odds and the best way to take advantage of them.

"Machiavelli was a game theorist," John Nash told us. So, too, was Vince Lombardi.

---

[*] In subsequent chapters we'll see how changes in the game in years to come—from new rules to a narrowing talent gap to bold new passing strategies—changed this payoff matrix, making passing a much smarter, and more widely used, strategy.

CHAPTER 5

DARWIN'S PLACEKICKER:
SURVIVAL OF THE FLATTEST

It all started with a funky freight elevator and an unspeakable accident. In the summer of 1941, Ben Agajanian was a promising placekicker and a 6 foot, 180-pound defensive end at the University of New Mexico in an era in which most players played more than one position. One day on his summer job at a Coca-Cola plant, he was riding a freight elevator, perched atop a barrel of the secret syrup. Agajanian carelessly dangled his left leg into the elevator shaft. But as the elevator climbed, Agajanian's foot was crushed by an unseen ledge protruding into the shaft. When he arrived at the hospital, the doctors focused on saving his life by stopping the bleeding. As for his foot, there was nothing to be done.

"What about football?" Agajanian asked.

The doctors told him that he'd be lucky to walk normally.

"Well, if you are going to have to amputate the toes, then square 'em off," Agajanian told the doctors. "Don't let one stick up. This way, I'll kick better." And that's just what the doctors did. With only his pinky toe remaining, Agajanian's right foot shrunk from a size 10 to a size 7.

Agajanian headed back to campus, hoping to resume his football career. His first thought was to kick left-footed, but that didn't work. He then tried kicking with his mangled right foot. Those first few times pain shot through his body when he struck the ball. When he took off his shoe, it was full of blood. Agajanian applied an over-the-counter remedy called Tuf-Skin to his mangled foot, but even then it was still too raw for kicking. He soaked his foot in brine—pickling it, in a way—and gradually the bleeding stopped and the pain gave way to numbness. Agajanian stuffed paper into the front of his old shoe where his toes had been, but that didn't work either. His coach, Ted Shipkey, sent him to a cobbler who built him a boot designed specially for placekicking. It was squared off with a strip of leather across the front.

"First one the guy made, it was crooked," Agajanian recalled. "I'd kick, and it'd go squirtin' off to the left. Second boot he made, he got it right."

That boot was more than just right. It's in the Pro Football Hall of Fame, because it helped Agajanian make history.

The man who would come to be known as "Bootin' Ben" quickly discovered that with his injured foot and his modified boot, he could actually kick farther and more accurately than he ever could with his toes. The accident and that boot helped Ben Agajanian secure a twenty-year career in professional football. And a special place in the evolution of football.

Kickers are football's ultimate outsiders. Today, they sit on the side-lines for most of the game, completely apart from the battles for yardage that dominate play. Despite the NFL's ongoing efforts to make the kicker's job harder and thus reduce the number of field goals, there isn't a player whose success—or failure—is reflected more directly on the scoreboard. In the 2012 season, New England kicker Stephen Gostkowski led the NFL with 153 points, or 9.6 per game. The leading scorer among non-kickers was Houston running back Arian Foster, who scored 102 points, just 6.4 points per game.

No position relies more heavily on the applied physics of perfect technique, and no position has evolved more radically as the game has changed. By examining the game's kicking specialists—from pioneers like Agajanian and Fred Bednarski to modern practitioners like Jim Breech—we can gain a deeper understanding of the game as a whole.

When the story began, football was a markedly different game. Under Walter Camp's nineteenth-century rules, a field goal was worth 5 points and a touchdown was worth only 2. Kickers were important members of the team, and Jim Thorpe was almost as well known for his kicking heroics as for running the football. But gradually, the game's powers-that-be came to realize that fans found touchdowns a more compelling product than field goals. They overhauled the rules; by 1912 a touchdown was worth twice as much as a field goal.

Harking back to the game's roots in rugby and soccer, every starter still played virtually every down on offense and defense. Offensive specialists also had regular—and usually corresponding—positions on defense, and vice versa. In 1943, Hall of Fame quarterback Sammy Baugh led the league in passes, punts, and interceptions.

In an environment dominated by generalists, placekickers became an afterthought, and the quality of kicking suffered. Kickers lacked strength, accuracy, or both, and teams all but stopped attempting field goals. In 1944, the Chicago Bears didn't attempt any field goals at all, and the Green Bay Packers won the championship that year without making a single field goal.

Paul Brown again came to the rescue. The Cleveland Browns' coach recognized an opportunity in the deterioration of the kicking game and once again revolutionized pro football. In 1946, he signed Lou Groza, a strapping 235-pound defensive end who would be best remembered as the modern game's first great kicker. In his rookie year, Groza set an NFL record with 13 field goals, and in 1953 he again broke the record with 23 field goals in 26 attempts. Groza gave the Browns a secret weapon, a way to salvage points from a stalled drive. Brown's tactical acumen paid dividends for the Browns, who emerged as perpetual championship contenders during Groza's career. In 1950, for example, Groza's 2 field goals provided the margin of victory in an 8–3 playoff win against the Giants in the American Conference title game, and his game winner with twenty-eight seconds left clinched the NFL championship with a 30–28 victory over the Rams.

Kicking matters. Especially when the other teams can't.

While that freight elevator accident ended Ben Agajanian's marginal prospects as an undersized defensive lineman, it opened a far more important window. He became the game's first full-time kicker.*

Joining the league around the same time as Groza, Agajanian started his career with the Philadelphia Eagles in 1945, playing a handful of games strictly on defense despite his damaged foot. Suiting up for the Steelers later that year, Agajanian broke his arm. The injury kept him out of the lineup, but Pittsburgh needed him to kick field goals and extra points, so the team sent him out with his arm taped to his chest. It soon became clear that Agajanian had an edge that no other kicker could match. With his damaged foot and the modified kicking shoe, his contact area with the ball was several times larger than that of a conventional toe kicker.

Agajanian was a perfect 4 for 4 for the rest of the season, and with that he took a giant step toward becoming the game's first kicking specialist. In 1947, Agajanian broke Groza's single season record

---

* Hall of Fame halfback Ken Strong did come out of retirement in 1944 at the age of thirty-eight to play for the Giants for a few seasons in which he just kicked field goals. But calling Strong a placekicker is a bit like calling Yogi Berra an outfielder because he patrolled left field and right field at the end of his career. The wartime signing of the popular Strong, who wore a wristwatch but no shoulder pads on the field, was more of a marketing move than a strategic one.

with 15 field goals. He played for the Los Angeles Dons strictly as a kicker and later hooked up with the New York Giants, where he was a member of their 1956 NFL Championship team, and was the kicker on the Packers 1961 Championship team. Respect for Agajanian's innovation was slow in coming: a number of times he quietly "retired" and headed back to run his profitable sporting goods business in Southern California, only to be lured back to the game by a team looking for an accurate kicker with a strong leg. Sometimes called the Toeless Wonder, Agajanian bounced from city to city, playing for eight teams in the NFL, the AFL, and the All-America Football Conference. He made 104 field goals and 343 extra points in a career that lasted until he was forty-three years old. With his deadly combination of accuracy and distance, Agajanian was valuable enough to his teams that they were willing to use a roster spot for him, even though he didn't play offense or defense. Most important, Agajanian's success helped to spur the growing trend toward specialization in the game. When he broke in, two-way players were the rule. By the time Bootin' Ben retired for good in 1964—for the fifth time—every player had one specific job on offense or defense, and the two-way player was a thing of the past.

Agajanian understood that he owed his remarkable career to that horrible accident. At one of his kicking camps, one of the fathers approached Agajanian and asked him earnestly, "How do I prepare my son to be a kicker?"

Agajanian paused as if deep in thought and said, "Cut off his toes."

But if you couldn't actually modify a placekicker's foot, you *could* modify the way it was used. Which is where Fred Bednarski and football's next generation of kickers enter the story.

"You kick the ball with the meat of your foot," explains Jim Breech, distilling the essence of the modern soccer-style kick into a single sentence. Like many kickers, Breech, who played for the Cincinnati Bengals and kicked what might have been the winning field goal in

Super Bowl XXIII— Joe Montana's last-second drive robbed him of that shot at immortality—is a great storyteller. He was a member of the first generation of young football players who dreamed about becoming a professional field-goal kicker when they grew up. A soccer-style kicker, to be specific.

"I was twelve years old in 1968, and I had never played any organized soccer," Breech recalls. Pete Gogolak had just become the NFL's first placekicker to approach the ball not straight on like Groza and Agajanian, but from the side, the way a soccer player might. "I thought that was a cool way to kick."

But what really sealed the deal for Breech? Two things: pain and kickball. Breech quickly learned that straight-on kicking is hard on the toes. "I couldn't get anything on the ball, and it hurt like crazy," says Breech.

"What spurred me on was playing kickball in the schoolyard," he continues. "I realized when I turned at a 45-degree angle I could get a ton of power. It was a natural motion, and it felt very athletic to me."

Breech took a kicking clinic with an instructor who knew how to teach straight-on kicking only. The coach grew frustrated with Breech's inability to get the ball up in the air. At the end of the day, the coach addressed the group and called Breech the most disappointing kid in the camp. In light of Breech's fourteen-year NFL career, that story has come to seem funny, but at the time those words stung.

And while Breech cites Gogolak as an early influence, he can't claim to be the very first soccer-style kicker. That distinction belongs to an almost totally forgotten player, a Holocaust survivor named Fred Bednarski.

Fred Bednarski grew up in eastern Poland, an area caught in the middle of the centuries of conflict between Germany and Russia. In 1939, Nazi German troops battled the Soviet Army for control of this strategically important real estate. When the Nazis gained control of the area, five-year-old Fred and his family were rounded up and put on a train to Austria. "They put us in these cattle cars, and they closed the doors," Bednarski recalls. The Bednarskis were taken to a labor camp outside of Salzburg, Austria, and the Nazis forced Fred's father to work in a radio factory. They lived in a communal barracks with one bathroom and, with food rationing, had very little to eat. "It was slow starvation," says Bednarski, noting that his mother's weight dropped below 80 pounds. The monotony was broken up only by the very real danger of Allied air raids. Bednarski recalls huddling in a bomb shelter for hours at a time. After World War II ended, the Allies liberated Fred and his family, but there was no home to return to. They moved to slightly better conditions in a displaced-persons camp, where young Fred was exposed to one of his lifelong passions: soccer. "We'd get some rags and tie them together and make a ball out of it," he recalls. "You had to kick the ball hard to get it away from the goal, and that's how I developed my kicking leg."

After the war, a displaced-person's lottery determined that the Bednarski family would move to Smithville, Texas, where his father would work on a dairy farm. The owner of the farm took young Fred to a high school football game, but the rules of the game were too complex. "I couldn't figure out what they were doing," he recalls. "I thought they were fighting."

At Fillmore Junior High, some kids were playing a game of pickup football, and they encouraged Bednarski to try kicking the ball. He walked over and examined the strange egg-shaped ball for a moment, and then approached it from the side, as if it were a soccer ball. He casually lofted it into low-earth orbit, 40 yards downfield.

"Who kicked that ball?" the high school coach asked later.

"The Polish kid, Coach," said one of the kids who was watching.

In high school, Bednarski became a kicker, a punter, and a linebacker, specializing in booming kickoffs. His friends suggested that he try to make the University of Texas football team as a walk-on. A muscular six-foot specimen (his younger brother, Joe, would gain fame as a professional wrestler under the name Ivan Putski), Fred soon earned a scholarship on the strength of those powerful kick-offs that often ended up as touchbacks. "It's like swinging a golf club," he says, explaining his revolutionary technique. "You have so much more control." But the highly restrictive—and short-lived—substitution rules of the day prevented Bednarski from becoming the team's regular field-goal kicker.

But they didn't prevent him from making history. On October 19, 1957, in Fayetteville, Arkansas, the Longhorns—under the leadership of new coach Darrell Royal—faced the tenth-ranked Razorbacks. In the first quarter, the Longhorns' drive stalled at the Arkansas 23-yard line. Royal gambled and told Bednarski to try a field goal, one of only 4 the team would attempt the whole season. Their opponents, it should be noted, hadn't tried any.

As Bednarski lined up to the side of the ball, Arkansas players began shouting "Fake! Fake!" wondering if it would be a pass or an end-around.

Bednarski calmly connected on a 40-yarder, the first soccer-style kick in college football history, the first points in an upset win for Texas.

The Arkansas players weren't the only ones who were baffled. "The officials had to talk it over to make sure it was legal," Bednarski recalls. "I was excited. I felt like I had contributed to the team. It's the kind of thing that can only happen in America."

Bednarski was, if anything, too far ahead of his time. It wasn't until a 2007 *Washington Times* story by Dan Daly that Bednarski got the appropriate credit for his historic kick, probably because most contemporary news stories referred to Bednarski as a "sidewinder" rather than as a "soccer-style" kicker. A couple of other largely anonymous soccer-style kickers—Evan Paoletti of Huron College and Walt Doleschal of Lafayette—followed Bednarski, but it wasn't until 1961 that Pete Gogolak of Cornell put soccer-style kicking on the map. Gogolak attracted attention a year earlier by kicking a 48-yard field goal in a freshman game, and on the varsity,

starred for an Ivy League team that attracted press coverage dispro-
portionate to its quality of play.

After Gogolak graduated in 1964, he tried out for the Buffalo
Bills of the American Football League. He kicked a 40-yarder. Then
a 45-yarder. Then a 50-yarder. After Gogolak connected from a near-
record 55 yards, Bills personnel director Harvey Johnson told Gogolak to stop kicking. "I didn't want to see any more, because I figured
he was going to cost us enough as it was," recalls Johnson.

As a rookie, Gogolak led the AFL in field goals and earned a spot
on the all-AFL team as Buffalo won the championship of the upstart
league. The next year Gogolak played without a contract, and in
1966 the New York Giants from the then-rival National Football
League saw Gogolak as the answer to the team's kicking problems;
they had made just 4 of 26 field goals in 1965 with kicker Bob Tim-
berlake missing 10 attempts in a row. Giants owner Wellington
Mara, who had found a place for Bootin' Ben Agajanian on his
championship teams of the 1950s, signed Gogolak to a three-year
contract for $96,000, an unheard-of salary for a kicker. This signing
would have repercussions well beyond the land of kickoffs and extra
points.

Until that time, there had been an uneasy détente between the
AFL and the NFL. The rival leagues fought tooth and nail over college recruits—to the point that they would each guard incoming
rookies with detectives to protect them from being poached. But the
NFL and the AFL had an unspoken agreement to not pursue each
other's veterans. Mara's signing of Gogolak broke that tenuous
peace, and the AFL countered by aggressively pursuing established
NFL players. That off-season, the league signed San Francisco's star
quarterback John Brodie as well as three players from Mara's Giants. The fear of an all-out bidding war provided the catalyst for the
resumption of the stalled merger talks between the AFL and NFL.
Within just a couple of weeks of Gogolak's contract, the two leagues
hammered out their historic agreement. All because of the battle for
pro football's first soccer-style placekicker.

———

"It's an amazing thing. It feels so easy," says Jim Breech. He's struggling a little, as athletes sometimes do, trying to fit a feeling into words, attempting to convey the sensation of kicking a 40-yard field goal that would be good from 50. "You know when you hit a golf ball dead center or hit a baseball on the sweet spot of the bat? You just don't feel it. It's just like that. It feels so easy."

But Breech, who came *this* close to being named Super Bowl MVP in 1989, knows better. There's a lot of hard work that comes before that Big Easy.

At the end of the 1988 season, Breech's Cincinnati Bengals made it to the big game, but the contest didn't follow the script. The Bengals featured their high-powered no-huddle offense, while the Niners sported one of the most potent attacks of all time. Without fullback Stanley Wilson, a former cocaine addict who relapsed only hours before the game, the Bengals' attack sputtered, but Cincinnati's defense rose to the occasion. So as the game settled into the fourth quarter, neither of those vaunted offenses had found the end zone. "It was a low-scoring game," Breech recalls.

As the Bengals drove into San Francisco territory once again with less than five minutes left in a tie game, Breech began his sideline ritual. "My normal routine is to watch the game. I love the game, and I've got a ringside seat," he explains. "Once we got over the 50, I'd be stretching and go over and start hitting some balls. I didn't have a set number. I kicked until I felt comfortable, until I had my swing down and I'd finish nice and balanced. It might be five or six, it might be seven or eight."

When the Bengals' drive stalled at the 22-yard line, Breech took the field. He started by marking a spot for his holder 8 yards behind the line of scrimmage. Breech would then move four steps back and three feet to the left of the ball to begin his approach.

As he set up, Breech remembered attempting several field goals from that same spot in practice the day before. It was then that he latched on to something that he found reassuring, given that 100 million people would be watching what he would do next. "I looked down, and there was my cleat mark from the day before."

Broken down into component parts, the next 1.3 seconds were very busy. As Breech approached the ball, he was beginning what Isaac Newton called a chain of momentum. In those four steps, Breech accelerated his body from a full stop to around 10 mph. When he stopped and planted his left foot on the cleat mark he found, he was transferring that momentum from his 190-pound body to his right leg. Because of Newton's conservation of momentum—remember, you can't destroy it, merely redirect it—Breech's right leg moved forward much faster than his body. The speed increased further as he snapped the joints of his leg forward in quick succession. Breech's knee was moving at around 15 mph, his ankle at about 33 mph, and his toe, about 40 mph. The duration of impact between Breech's foot and the 0.91-pound ball was just 0.008 seconds, but before it reached the goalpost, the ball attained a speed of 70 mph.

In that hundredth of a second, as Breech's size-5 Adidas Copa soccer cleat made contact with the football, the outcome of a Super Bowl hung in the balance. What's really going on at that crucial moment? Plenty. And it's the moment in which Breech and his soccer-style brethren channel Ben Agajanian.

"You hit the ball just below the center with the meaty part of your foot," Breech explains. "You put your laces on the ball." Breech's goal is to achieve as much contact area between his foot and the ball as possible, which, again, is why Agajanian had an edge over the toe kickers of his day.*

A soccer-style kicker gets this advantage naturally, by turning his

---

* As did Tom Dempsey of the New Orleans Saints, who was born with a congenital defect of his right foot, and like Agajanian kicked in a straight-ahead style wearing a shoe with more surface area. In 1970, Dempsey converted an NFL-record 63-yard field goal, a mark that still stands although three other kickers have since tied Dempsey's feat. That prompted the NFL to adopt what became known as the Tom Dempsey Rule, which prevented kickers from using special shoes with a larger striking surface.

foot sideways and making contact with his entire instep. Why is a larger contact area better? For exactly the same reason that race cars have wide tires. A narrow tire will slip and skid during heavy acceleration because the power from the engine overwhelms the friction between the tire and the road. While the forces involved in field-goal kicking aren't nearly as large as a 700-hp turbocharged V10 engine, the larger contact area still allows for a more solid connection between foot and ball at the point of impact.

Kickers will go to extremes to enhance this contact area. Breech, for example, wore a size-7 cleat on his left foot but squeezed his kicking foot into a size-5 soccer shoe, two full sizes smaller, for better foot-to-ball energy transmission. Some kickers, like Philadelphia's Tony Franklin, who entered the league in 1979, took it a step further, actually kicking barefoot. Franklin would apply a little bit of tape to his foot to avoid what he called "Christmas toe": His skin would turn green from rubbing on the turf, and red from blood.

"It's all about feel," says Breech.

The foot is only half of the equation, however, and there's no one fussier about the condition of a football than an NFL placekicker. A new football is hard and slippery and difficult to kick. Breech likens it to trying to play in a pair of new penny loafers instead of your worn-in sneakers. Kickers once went to great lengths to avoid that, both legally and illegally.

They'd start by brushing the ball to break up that synthetic coating and inflating it above the conventional 13 psi of pressure to stretch it, then deflating it so that it could be manhandled. Some kickers took this last part to extremes.

"Footballs have been steam-bathed, baked in aluminum foil, dunked in water, brushed with wire, bonked with hammers, buffed with strips of artificial turf, jumped on, shot out of Jugs machines, pounded into walls or racquetball courts, inflated and deflated more often than Oprah Winfrey, Armor All–ed, shoe-polished and lemon-aded, crushed under weight-lifting plates, and like a female wrestler at a county fair, dunked in evaporated milk," reported Jack McCallum of *Sports Illustrated* in 1999. "Maybe even microwaved." Over the years, NFL kickers employed some or all of the measures mentioned above to make a ball softer and rounder. And, not coincidentally, more like a soccer ball. In 1999, the NFL tried to stop these shenanigans by introducing the K-ball. These balls, which are used only on kicking plays, are branded with a "K" at the factory and shipped straight from the manufacturer in a sealed box that's opened only two hours before the game. The ball that Breech kicked that evening in Miami? Kickers can't mess with the ball in the Super Bowl.

Kicking the ball, that's the easy part.

"It's not just being able to show a skill," says Sian Beilock, a University of Chicago cognitive researcher and the author of *Choke: What the Secrets of the Brain Reveal About Getting It Right When You Have To*, "but to show it in stressful situations."

Kicking a football is a reasonably straightforward act, with relatively few variables. Except when those 1.3 seconds will result in either modest success—kickers are *supposed* to make 40-yard field goals—or life-altering failure. In 1991, Scott Norwood missed a field goal that would have won the Super Bowl for the Buffalo Bills. In Vincent Gallo's 1998 movie *Buffalo 66*, the lead character, Billy, plots to kill a thinly disguised Scott Norwood doppelgänger because the missed kick caused him to lose a $10,000 bet that landed him in prison. The villain of *Ace Ventura: Pet Detective* is a Norwood-like kicker who missed a potential game winner in the Super Bowl and went criminally insane. They don't make movies about Adam Vinatieri, who made two Super Bowl–winning field goals.

"We've all known moments of failure, of blowing an exam, of being fired, spurned, disgraced, yet these moments are seldom public," wrote Karl Taro Greenfield in a *Sports Illustrated* profile of Norwood. "How do you go on when every time you walk into a liquor store or a gas station, there is someone pointing at you, reminding you of your worst moment?"

That's what Jim Breech was trying to avoid on January 22, 1989.

For all the tweaking of shoes and massaging of footballs, the thing a kicker needs to control most is his mind. Beilock notes that when a field-goal kicker faces a tense situation, it activates the same part of the brain—a pain pathway in the insula—that would be stimulated if he were stuck with a needle. "Our neural real estate doesn't always make real distinctions between what's psychological and what's physical," says Beilock, "which brings added power to this idea that pressure can have a detrimental effect."

In other words, pressure is real.

Long before he ever thought about the Super Bowl, Breech devised a mental routine that was every bit as crucial to his success as his physical preparations. "That routine is so critical," Breech explains. "If you're in a high-pressure situation you don't want to spend a lot of time thinking about it. You allow the routine to take over."

For Breech, it began with the third goalpost. The actual goalpost defines the outer boundaries for a good kick; if the ball goes anywhere near the upright, it's too close for comfort. Humans are naturally wired to aim where we're looking, which is why YouTube is full of videos of drifting drivers plowing into the only light post in an otherwise empty parking lot. Watching the actual goalpost is a recipe for failure. So Breech created an imaginary third upright that would sit between the two real uprights.

"I would see this in my mind's eye, and I'd move it a little to compensate for the wind," he recalls. Since the wind was blowing from his right to his left on Super Bowl Sunday 1989, Breech slid the imaginary goalpost a little to the right of center.

What else was rattling around in Breech's brain? Magic words. "I always had a key word or two that I'd focus on," he explains. The two words that day: *tempo* and *balance*. Tempo addressed Breech's speed in approaching the ball, but it's telling that he used a musical term. There is no perfect speed for a kick. Some kickers, like Detroit's Eddie Murray, approached the ball very quickly, while others, like Breech, needed to be somewhat more deliberate. By focusing on tempo, Breech was thinking about staying in sync with his snapper and his holder and his own internal metronome.

"If you're too quick, you might jump at the ball and yank it left," he explains. And in a pressure situation, the temptation to speed up is great. The day before the Super Bowl, Breech did an interview with the Kansas City Chiefs' future Hall of Fame kicker Jan Stenerud, and afterward, they did what kickers do: talk about kicking.

"The one thing about these games is everything speeds up," Stenerud told Breech. "It feels like you're going 100 miles per hour. So if you get into a situation where there's a big kick, really . . . slow . . . yourself . . . down."

From Beilock's perspective, a mantra like this performs another overarching function. "It's a part of a psychological toolbox. We can say, 'Don't overthink this,' but then we go and do just that," she explains. She notes that by the time Breech was playing in the Super Bowl, his kicking motion was already part of an unconscious sub-routine. He could run on autopilot. The job of his conscious mind was to stay out of the way. "When the goal is to execute fine motor movements, that extra attention can be counterproductive. It messes up the subroutines," she explains. "If you're shuffling down the stairs, and I ask you to think about what you're doing with your knee, there's a good chance you'll fall on your face." The way around that is to give the conscious mind some busywork that will prevent it from overanalyzing the task at hand. "A mantra," according to Beilock, "prevents you from breaking it down into these components that muck you up."

Breech's other internal cue was balance. The end result of this one was more obvious to the fan. Breech describes a smooth motion in which he moves smoothly through the kick and his body continued toward the end zone as he lofted the ball toward the goalposts.

"I know if a field goal is good as soon as I kick it," Breech explains. "If I'm in balance at the end of the kick, I know I nailed it. If

I'm wobbling a little, then I'm watching it like everyone else in the stands, hoping it makes it through."

Beilock views Breech's balance prompt as priming the pump, in a way. "I love the 'outcome' example. When you're doing something that's pretty well practiced, focusing [your mind] on the end result gives your motor system the end result," she says. "It essentially tells your body what to do."

But strangely, as Jim Breech was about to kick the biggest field goal of his life, he knew that what got him here was, well, *failure*. He still recalls missing a kick at Cal that would have beaten USC and might have earned the team a spot in the Rose Bowl. "I didn't feel nervous," he recalls. "I just missed the kick. I pulled it left." And while competing for a job with the Raiders he missed several important kicks and was cut soon after.

While he was sitting at home waiting for another NFL team to call, he came to a simple but powerful realization that would change his outlook forever: "What is the deal here? Why am I so worried about all the things I can't control? What are people going to say? What's going to be written about me? What are my teammates thinking? I realized all I can control is the kick. That's it. It allowed me to focus in a lot more closely on what I was doing versus having these things run around all over my head. All of a sudden a lightbulb went on, and from then on I did really well on those kind of kicks."

And with that, Jim Breech put aside his teammates and the newspaper columnists, the stadium full of fans and a nation full of television viewers. He looked down at the cleat mark from the day before.

"Tempo," he said to himself. "Balance."

He took one step toward the ball, then another, and another.

"It wasn't the greatest kick in the world. Normally you have a pretty good feel about it. That one, I thought it was going to be okay. It was about right center and it came back in toward the middle," Breech recalls. "But you're always happy when the officials put up their arms."

As he headed toward the sidelines where his teammates were cel-ebrating, the assembled media was doing its jobs, too. Reporters had filled out a preliminary ballot for the game's MVP. With 3 field goals and an extra point to his credit, Jim Breech had earned the trophy and a shiny new Corvette. Unless . . .

Breech saw his teammate, wide receiver Cris Collinsworth, on the sideline. He gave Breech a high five, and Breech grinned back, know-ing that he had overcome the physical and psychological challenges to take care of the one thing in his control.

"Great kick," Collinsworth told his old friend. "But there's too much time left for Joe Montana."

PART TWO

THE GAME

"If a single flap of a butterfly's wings can be instrumental in generating a tornado, so also can all the previous and subsequent flaps of its wings, as can the flaps of the wings of millions of other butterflies, not to mention the activities of innumerable more powerful creatures, including our own species."

—EDWARD LORENZ

THE BUTTERFLY EFFECT OF GREG COOK

In the baseball movie *Bull Durham*, veteran catcher Crash Davis says to phenom pitcher Ebby Calvin "Nuke" LaLoosh, "You've got a gift. When you were a baby, the gods reached down and turned your right arm into a thunderbolt."

Greg Cook had one of those arms.

Tall and blond and model-handsome, the 6'4", 220-pound Cook was the archetype of a professional quarterback. He was big, strong, and fast, and he could throw the ball a proverbial country mile. Playing for the University of Cincinnati, he set an NCAA single-game record by passing for 554 yards. In 1969, Cook was drafted by his hometown Bengals in the first round with the fifth overall pick. He was an overnight sensation. At that point in NFL history, rookie quarterbacks, even highly touted ones like Cook, rode the bench and waited their turn. Not Cook. He grabbed the starting job and proceeded to carry the Bengals on his back.

His arm was remarkably powerful—Cook claimed, matter-offactly, that he could throw the ball 90 yards on the fly—but it was more than that. He supplemented his strength with a soft touch and enough mobility to allow him to buy time for his receivers to run those long routes.

Bengals owner Mike Brown called him "John Elway before John Elway." His teammate Ken Riley compared him to Joe Namath: "We won games because of the way Greg Cook could throw the ball. He had that quick release, and he was confident."

"Cook," said Bill Walsh, the Cincinnati Bengals coach who would go on to a Hall of Fame career with the 49ers, "may have been the greatest single talent to play the game." This coming from a man who coached Joe Montana, Steve Young, and Dan Fouts.

In his second professional game, Cook went 14 for 22 for 327 yards with three touchdowns in Cincinnati's 34–20 victory over the Chargers. The second-year expansion team won its first three games, not in spite of the young quarterback but *because* of him.

"There has never been an NFL rookie like him," wrote Paul Zimmerman in *Sports Illustrated*. "And almost everyone who saw him play has some personal memory—the 70-yarder he threw to Bob Trumpy, the 60-yarder to Eric Crabtree, the deep passes that came off his arm like rockets, but rockets delivered with perfect touch and timing. They remembered his poise and savvy and instinctive knowledge of how to attack a defense."

In his rookie year, he averaged 17.5 yards per completion. It was an NFL record that has never been equaled. Cook also led the league in completion percentage, yards per attempt, and passer rating, and won the UPI's NFL Rookie of the Year award.

Much of the credit for Cook's quick rise belonged to Walsh, his coach. The legendary Paul Brown was the head coach of the Bengals at the time, but he ceded responsibility for the offense to Walsh, who was technically the receiver's coach but who functioned more like a modern offensive coordinator. Walsh designed an offense around Cook's otherworldly arm and his remarkable athleticism. He had coached in Oakland and built an attack in Cincinnati that was based on the Raiders' go-deep philosophy but specially tailored to Cook's remarkable skills.

"His philosophy was based on stretching the field, which would force the linebackers deeper and open things up underneath. Then he'd go deep again," said Cook, discussing Walsh's offense in a 2001 interview. "He always liked deep receivers. He liked to force the cornerbacks downfield, then go short to bring 'em up, then go deep again. It was like the horse on the merry-go-round, up and down, up and down. With the [defensive backs], it was up and back, up and back. It was merciless. He had people worn out by halftime. By the end of the half, they didn't know what they were doing," Cook explained. "It was never a take-what-they-give-us philosophy. It was make-them-take-what-*we*-give-*them*. And it gave me a feeling of invincibility. I felt I could make any throw he wanted me to make."

Chances are, however, that unless you're a longtime Cincinnati Bengals fan, you've never heard of Greg Cook. And that's because Cook's promising career ended almost as suddenly as it began.

Midway through only his third game, against the Kansas City Chiefs, Cook was sacked by Jim Lynch. It wasn't a spectacular hit and Cook walked off the field casually, having injured his throwing shoulder. In 1969, sports medicine was in its infancy, and after X-rays revealed no broken bones, the team assumed that Cook's arm would heal quickly. Cook had actually torn his rotator cuff and a biceps tendon. If that had happened today, he would have been shut down for the season, undergone surgery, and after a few months of rehab, come back as good as new the next year. Instead, Cook missed a few games and, as any cocky twenty-one-year-old might, begged to get back into the lineup. "I took cortisone shots and played in pain," Cook recalled. And played brilliantly, even at less than full strength.

By the off-season, however, Cook couldn't lift his arm. "It got to the point that I could lie on the side of the bed and hang my arm off it, and when I'd try to lift it, the shoulder would go out," Cook told Zimmerman. "I'd get sick to my stomach. What's going on?"

Cook finally had surgery, and he would miss almost three full seasons.* It was too little, too late. He played in one game for the Bengals in 1973, then signed with the Chiefs and was out of football in 1974. Cook's NFL career lasted a total of twelve games, but he played healthy for only ten quarters. He died of pneumonia on January 27, 2012, at the age of sixty-five.

"He was the prince who never became king," said Brown. What wouldn't become clear for years is that Cook's hard luck changed the face of the NFL forever.

Edward Lorenz was a meteorologist at MIT. In the winter of 1961, he had constructed a basic mathematical model for global weather systems accounting for factors like wind speed, humidity, and temperature differentials to predict weather patterns. He programmed a primitive computer to run his calculations.

During one of his early trials, Lorenz wanted to use a bigger sample. But instead of running the entire program again from the beginning, which would have taken hours, Lorenz started it in the middle. He entered a value into the computer from a printout of his previous work and went out to have a cup of coffee. When he returned, he examined the results, and what he saw baffled him.

The end result was wildly different from that previous, seemingly identical run of the program. At first he thought the computer had malfunctioned, perhaps blown a vacuum tube or something. But Lorenz looked carefully and concluded that the computer was working fine.

These two vastly different results were the result of a tiny change in his own input. On the original program, the computer stored the number in question out to six digits: 0.506127. On the printout that Lorenz used for the second run, that number was rounded off, so instead he punched in a three-digit version: 0.506.

That difference of 0.000127 had changed everything.

* A contemporary newspaper account of the surgery, probably relying on misinformation from the team, reported that Cook had "hurt his arm in a pickup basketball game last winter."

At the time, this wild variance wasn't of interest to anyone except Lorenz, but a decade later, almost on a whim, Lorenz presented a paper that codified his ideas. Lorenz likened the minuscule differ-ence between those two values to that of a butterfly flapping its wings. And he posited that some systems—like the weather—are so complex that an event as small and random as an insect fluttering in Brazil could determine whether a tornado forms in Texas.

Or, perhaps, whether a quarterback's injury in Cincinnati would, a decade later, lead to a groundbreaking offense in San Francisco.

Lorenz called it the Butterfly Effect.

If Greg Cook looked like the Hollywood conception of an All-American quarterback, Virgil Carter looked like, well, a math pro-fessor. Desperate for a quarterback after Cook's injury, the Bengals traded a sixth-round pick to the Chicago Bears for Carter, who had just lost his job to the strapping Bobby Douglas. At 6'1" and only 190 pounds, Carter was not a particularly imposing physical speci-men. If Cook had perhaps the best arm in the league, Carter's was decidedly below average. Cook could haul off and hit a receiver some 60 yards away, while Carter couldn't throw more than 20 yards with much accuracy. But whatever Carter lacked in arm strength, he made up for with a very quick mind. In the off-season, he actually did teach math at Xavier University.

Walsh immediately recognized that his new quarterback would require a new offense. So he took the system that he designed for Cook and rotated it 90 degrees. The Cook offense was based on deep, vertical routes that would stretch the defense the length of the field. Walsh had learned this AFL-style passing attack from Al Davis during his stint as an assistant with the Oakland Raiders, with Davis having learned it from Chargers coach Sid Gillman. But Carter, like most NFL quarterbacks at the time, didn't have the arm strength to make those long throws. So Walsh designed a passing game based on an idea that, at the time, seemed strange. Instead of running pass routes down the field, Walsh's offense sent receivers moving *across* the field. Wholly counterintuitive, it was an idea that would ultimately revolutionize football.

At that point, even with a quarterback as talented as Cook, the pass was a secondary option. The deep pass was a low-percentage play, one that a conservative coach like Paul Brown—or Vince Lombardi a decade before—would use to keep the defense honest. Teams that could run did run. The Bengals were an expansion team and didn't have the talent to execute the kind of running attack that Brown would have preferred, so they tried Walsh's unconventional passing attack out of sheer desperation.

In Walsh's new scheme, the pass would morph into a new type of weapon. Walsh took the timing patterns developed by Gillman and elevated them to a new level. When the timing between Carter and receivers like Bob Trumpy clicked, it was a thing of beauty. More important, these short, precise passes succeeded quite often. Carter would lead the NFL in pass-completion percentage in 1971, completing 62.2 percent of his passes, more than 10 points higher than his career best. Walsh's offense also reduced the chances of defenders picking off passes; Carter's interception percentage dropped from 5.2 percent in Chicago to 3.4 percent in Cincinnati.

The biggest benefit of Walsh's new scheme was that the catch was only the beginning. By catching the ball in stride, the receiver could take a 10-yard pass and turn it into a 25-yard gain, with the yardage he tacked on by running with the ball after the catch. The weak-armed Carter completed the longest pass of the 1971 season, a 90-yard touchdown to Speedy Thomas, 76 yards of which came after the catch. In essence, the pass became a long handoff. And it was a handoff that took place *behind* the defensive line.

In Carter's first year, the Bengals improved markedly. They went 8–6, winning the AFC Central and making the playoffs in the franchise's third season. The Bengals drafted Ken Anderson out of tiny Augustana College. Anderson shared Carter's mental acuity, but with a bigger body and a better arm. In 1974, Anderson set a franchise record with a 64.9 completion percentage and established an

NFL single-game record in that category, when he went 20 for 22 against Pittsburgh's formidable defense. In 1974 and 1975, he led the NFL in yards per game and the all-important yards-per-attempt stat.

But it didn't work out the way it should have for Walsh in Cincinnati. Although Brown was a coaching legend and a brilliant innovator in his own right, he was also something of a control freak. He would ask Walsh—positioned up in the press box—to radio him a passing play and have line coach Tiger Johnson on the sidelines propose a running play. Whenever possible, Brown would choose the running play.

Despite the team's success, Brown, who was the team president as well as the coach, didn't trust Walsh. When it came time for Brown to retire, after an 11–3 season in 1975, instead of promoting his heir apparent Walsh, Brown gave the job to Johnson. To add insult to injury, Brown had the team make the announcement when both he and Johnson were out of town, so that Walsh was left to face the media and put a positive spin on being passed over for the promotion. Walsh left Cincinnati, spending one season as an assistant with the Chargers—where he tutored young passer Dan Fouts in the Gillman-style passing attack—and then took the head coaching job at Stanford University. Meanwhile, whenever an NFL head coaching position would open up, the influential Brown would quietly bad-mouth Walsh to his fellow GMs. One of the game's greatest innovators was being blackballed.

Which is why Walsh's revolutionary new offense, which had helped to propel a bad team to surprising success, would have to wait another half-decade to get a full trial. But if Greg Cook had taken a step to his right to elude that tackle against the Chiefs, it's not hard to imagine an alternate reality in which Paul Brown learns to trust the pass and names Walsh to succeed him as coach of the high-flying Bengals. And football is never reinvented.

Instead, Cook's injury was football's flap of a butterfly's wing.

CHAPTER 7

AS MEL BLOUNT CHANNELS
THOMAS EDISON

On August 12, 1978, sports agent Jack Sands negotiated a contract extension that would make New England Patriots wideout Darryl Stingley one of the highest-paid receivers in the National Football League. The deal would be announced the following week, so before the meaningless preseason game against Oakland that would be played that Saturday, Sands warned Stingley half in jest, "Don't sprain your ankle."

In the early stages of that road game, with Stingley and starting quarterback Steve Grogan still in the game, New England had driven deep into Oakland territory. They had the ball, 3rd and 8, on the 24-yard line. Grogan called a play known as 94-Slant.

Stingley would be the primary receiver, lining up on the strong side, near the tight end. He would head downfield for 8 yards and then cut across the middle at a 45-degree angle. Stingley was open for a split second in front of cornerback Lester Hayes, but Grogan's pass sailed high over Stingley's head.

In a 1983 issue of *Ebony*, Stingley wrote about the hit that followed:

> I never had a chance to catch the pass. The ball just flew past my outstretched fingertips as I leaped into the air as high as I could. The ball was so close I could feel it going by. I was coming back to earth, and what did I see, suddenly in a flash—I can still see it now plain as day: I'll never forget it—Jack Tatum, number 32, barreling toward me. I felt as though I was suspended in midair, a feather succumbing ever so slowly to the pull of gravity, and here was this monster train coming at me full steam.
>
> He was 210 pounds. I saw him coming and dropped my head to get it as low as possible so I could duck. But it was too late. He delivered the blow. He cracked me on the head and on the back of my neck with full force. I hit the ground with a thud and tried to get up as I had so many times before, but I couldn't move.
>
> I felt like I was the cornerstone of a high-rise building. As if an elephant had his foot on my chest. I wasn't in pain or at least I couldn't feel any. I just couldn't move. Not a muscle. Nothing. I couldn't feel my feet. Or my arms. Or my body. I couldn't feel anything.

The grainy video of the play shows Stingley extended and reaching for the ball, and before he gets his feet back under him, Oakland defensive back Jack Tatum comes from Stingley's right and hits him just under the chin with his shoulder pad. After the play, Stingley slumps to the turf, and the medical staff can be seen checking his extremities. The hit fractured Stingley's fourth and fifth cervical vertebrae and compressed his spinal cord. Stingley was paralyzed from the chest down and came perilously close to dying from respiratory complications over the next few days. He remained a quadriplegic until his death in 2007.

Tatum had a reputation as one of the game's hardest hitters, and the Raiders of the 1970s were well known for intimidating their opponents with their rough play. But Tatum's hit on Stingley was legal, and no penalty was called on the field or imposed after the fact. The impact of this tragic play was such that it didn't matter.

"I saw replays many, many times, and many times Jack Tatum was criticized. But there wasn't anything at the time that was illegal about that play," said New England coach Chuck Fairbanks. Legal or not, however, the play marked a profound change in the rules in the NFL—and their enforcement. "I do think probably that play was a forerunner for some of the changes in rules that exist today that are more protective of receivers, especially if there is head-to-head-type contact."

Major League Baseball had its own dark moment. On August 16, 1920, the Cleveland Indians were playing the New York Yankees at the Polo Grounds on a dreary afternoon. In the fourth inning, Yankee pitcher Carl Mays threw a high inside pitch to Indian shortstop Ray Chapman using his unorthodox sidearm style, which one contemporary called "a cross between an octopus and a bowler."

Whether it was the bad lighting or Mays's unconventional delivery, Chapman simply didn't see the pitch. As the ball approached, he didn't flinch, and it hit him square in the head. The sound of the ball hitting Chapman's helmet-less head was so loud that Mays, thinking the ball had hit the bat, instinctively picked up the ball and threw it to first. Chapman collapsed and was rushed to the hospital. He died the next day, becoming the first and only major leaguer to be killed by a pitch during a game.

If Chapman's death had led to a eulogy and nothing more, then it would be merely a sad footnote. But instead, baseball used the tragedy as a springboard for a variety of rule changes that would transform the game, in ways predictable and not. There's a direct line to be drawn from the tragic figure of Ray Chapman to the emergence of Babe Ruth. A similar line can be drawn from the tragedy of Darryl Stingley to the rise of Joe Montana.

Charles Darwin wouldn't have missed a rules committee meeting. When a sport changes its rules, a similar evolution on the field can't be too far behind.

And in this way, it's not unlike biological evolution in all its nuanced glory. When most non-scientists think about evolution, they fixate on the seemingly miraculous transformation that an animal species undergoes. A short-beaked finch turns into a longer-beaked finch. A brown toad turns into a bright green one. An orangutan turns into Louis CK.

These transformations, or the absence of them, are what drives the sociopolitical arguments about evolution. That's what Clarence Darrow argued in the Scopes Monkey Trial. But while there's much energy expended over the "how" of evolution, what gets swept under the rug is the "why." The reason why plants and animals evolve is at once simple and complex: their environment changes. Some of these are long-term changes that happen at glacial speed, like the ones Darwin posited. Others happen quickly, with a speed that would have surprised even Darwin. The evolution that accompanies them can be just as quick and surprising.

"Darwin says in *The Origin of Species* that evolution takes place over geological ages—that we never see it in our lifetimes," explains Jonathan Weiner, author of the Pulitzer Prize–winning book *The Beak of the Finch*. "But in fact you can watch it, and you know it's evolution by natural selection if you're watching carefully enough."

The husband-and-wife team of researchers depicted in Weiner's book have been doing just that. Since 1973 Peter and Rosemary Grant have returned annually to a small island in the Galapagos called Daphne Major. They have documented the size and beak configuration of the Darwin's finches that inhabit the island, along with the changes in the birds' environment that prompted the evolution of the birds' beaks.

In 1977, for example, a severe drought inhibited plant growth on the island. The "low-hanging fruit" of the local food supply—small, soft seeds—quickly disappeared. What remained were larger, harder seeds that, under normal conditions, the birds would ignore. The larger birds with strong beaks adapted to these conditions by consuming this alternate food supply.

"What's left are the big tough seeds that are hard for those birds to crack open. It is only the birds with the nutcracker beaks that are able to get them open, so they survived the drought," says Weiner. "The smaller ones with the weaker beaks don't make it."

The next generation of finches—the offspring of the survivors of that difficult season—tended to be larger birds with more robust beaks.

"They have the bigger, deeper beaks and they inherit the earth," Weiner says.

Seven years later, the evolutionary pendulum on the Galapagos swung in the other direction. A particularly wet growing season altered the food supply again and with it the birds' foraging strategy. "The birds with the massive beaks are at a disadvantage in a wet year, because the island gets covered over with grasses and vines and they can't get at the big tough seeds that their beaks are best at cracking," Weiner explains. "It is the smaller individuals with the smaller, pointier beaks that are able to get the little grass seeds and crack those, and they make it through."

Weiner claims that, although he'd been working on the book for a year, he didn't really understand the process on a gut level until he went to the Galapagos and saw this life-or-death competition play out before his eyes.

"These scruffy little birds, they are struggling. The cupboard is bare," he recalls. "That is a dramatic scene. It is grim. So then the lesson sinks in."

At one level, evolution itself can be properly described as change. But, as Weiner and the Grants have chronicled, evolution is also *driven* by change.

"The world changes and favors those organisms that can change with it," says Weiner. "That's really what evolution is about. That's what it comes down to. The world is a turbulent place. It is always changing, on time frames from minutes to millennia. As the environment changes, we evolve generation after generation to fit the environment we find ourselves in."

The environment of a football game may be a man-made construct, but it's an environment nonetheless. When you see a fundamental shift in the way the game is played, rest assured that the alteration came about as a result of an underlying change in that environment. Those changes can come in the form of a unique player or a bold new strategy, but either way, the effects tend to spread slowly and fitfully. Not so when the evolution is being driven by rule changes. Again and again, from the days of Walter Camp and Teddy Roosevelt to the modern NFL Competition Committee, the league's movers and shakers would meet and update the rules. Unlike changes wrought by a team's strategy or a single player, rule changes are universal—they apply equally to every team—and often, like a rainy season in the Galapagos, they crop up suddenly. Coaches and players and teams are forced to adapt to them, often in unexpected ways.

And so it was at the start of the 1978 season, according to Miami Dolphin coach Don Shula, a longtime member of the NFL's influential Competition Committee. Their mandate as they met that year? "To open the game up and make it as exciting as we could make it," Shula recalls. "And make it as safe as we could make it."

The 1960s and 1970s were dominated by teams like Lombardi's Packers, Shula's Dolphins, and Chuck Noll's Steelers, that played boring smashmouth football dominated by aggressive defenses, conservative run-based offenses, and low-scoring games. The committee called for an unprecedented makeover of the rules to encourage more passing.

"We wanted more big plays," Shula recalls. "Instead of three yards and a cloud of dust, we wanted to make it fifteen, twenty yards. Spectacular plays. Receivers and defensive backs going high in the air making great plays. Running after the catch. All of those things that opened the game up."

Shula's rules package included changes that liberalized blocking and reined in pass defense. Which rule change had the most impact?

"I think the one that said you could only jam the receiver in a five-yard area as he left the line," Shula argues. "Then after that, you couldn't hit him."

Also known as the Mel Blount Rule.

Is there a more singular honor in the world of sports than having a rule named after you? When the league singles you out, it suggests that some part of your game sat right on the ragged edge, living in the land of the loophole. And that was Mel Blount's claim to fame.

Blount was a Hall of Fame defensive back for the Steelers of the 1970s, one of the key players on a team that won four Super Bowls and is universally considered one of the sport's greatest dynasties. The team's hallmark was their Steel Curtain defense, and while players like "Mean" Joe Greene grabbed the spotlight, it was Blount who changed the game.

Until Blount came along, most defensive backs approached the game in a reactive manner. Defensively, as it were. They shadowed the receiver, and only when the ball was thrown did they go after him. Blount put a certain measure of *offense* into *defense*.

Blount was 6'3" and 220 pounds, a good 25 pounds heavier than most players at his position. He built his career on the perfectly legal strategy of playing bump-and-run defense, hitting the receiver at the line of scrimmage before dropping back into coverage. "He was one of the first big cornerbacks," explains Bengals quarterback Ken Anderson, who faced Blount regularly in the AFC Central. "He was fast enough to cover anybody you put out there.

"If he came up and pressed you, with the rules the way they were, he could literally mug you all the way down the field. It was hard for receivers to get open," Anderson continues. "If you looked at the scoring averages back in those days, the point totals were lower. If you threw for 2,700 yards you might have led the league in passing. If you threw for nineteen TD passes you might have led the league too. The powers that be in the NFL wanted more scoring, and that's when the rules came in that opened up the passing."

"[Blount] was a big, strong defensive back," Shula adds. "And when he'd take on a receiver, there was nowhere to throw the ball. He was one of the guys we talked about for the techniques he used."

It's not that Blount broke the rules. And it's certainly not that he was a dirty player. It's simply that most players model their game on what their coaches teach them or what they watch their teammates and opponents do. They have only a vague sense of what's possible. Not Mel Blount. He would find a way to slow down a receiver, and if the official didn't throw a flag, he'd try it again. If he got away with it again, he'd put that tactic in his toolbox. And eventually, piece by piece, Blount built a game. A Hall of Fame game.

And in that way, Blount channeled Thomas Edison. Science is a bit like walking in a dark room with a dim flashlight looking for the exit. If you walk into a wall, that may seem like a failure to an outside observer, but you've actually acquired a valuable data point. Whack the wall, and you know where the door *isn't*. Edison's creativity was based on this type of pragmatic experimentation. When he invented the lightbulb, he tried not dozens, not hundreds, but *ten thousand* different materials as potential filaments, ranging from bamboo to platinum, from hickory to human hair. He didn't spend much time pondering which ones would work and which ones wouldn't. He just tried them. Modern consultants call this divergent thinking.

To speed up this process, Edison assembled his own army of scientists and technicians called, colorfully, the Muckers.* Edison would tell the Muckers what to try, and they'd try it. When it didn't work—as most of the trials didn't—Edison would shrug and cross that possibility off his list. When it did work, he would pat himself on the back and take credit for another successful project. He would ultimately encourage his Muckers to employ the same trial-and-error approach. In the early days of the lightbulb the biggest problem was securing the bulb to the fixture, and the Muckers tried any number of potential solutions. One member of Edison's army looked at a kerosene bottle and wondered if its threaded cap would work on a bulb. He tried it. It worked. And since that same threading is now used in CFC and LED bulbs, this impromptu solution is poised to outlast the incandescent bulb.

Even outright failures often yielded unanticipated benefits. The Muckers, needing insulation for a transatlantic cable, developed a carbon putty. The material didn't work for that cable, but it was a key component in the compact microphone that would make the telephone feasible. "The real measure of success is the number of experiments that can be crowded into twenty-four hours," Edison said. And that seemingly random and haphazard process is the reason that Edison accumulated 1,093 patents—as well as more than 500 unsuccessful patent applications. That's a patent application every week for thirty years.

"To invent, you need a good imagination and a pile of junk," Edison argued. For Mel Blount, getting called for a pass interference penalty was part of that pile of junk.

A process that ended with a rule with Blount's name on it.

"They always legislate in favor of the offense," Blount said wistfully after the rule change. "Defense may win championships, but the rules always favor the offense."

---

* The name came from one of their early ventures, a project to develop bricks that wouldn't absorb moisture, in which a year was spent experimenting with different binding agents, or "muck."

When Darryl Stingley suffered his tragic injury, the Mel Blount Rule was already on the books. It bears repeating that Tatum's hit was legal under both the old and the new rules. So it was a combination of new rules and more stringent enforcement that changed the environment of the NFL so profoundly in the seasons after Stingley's injury.

That's exactly what happened in baseball in the early 1920s. During the first two decades of the twentieth century, pitchers routinely threw a pitch called the spitball, in which they applied tobacco-stained saliva, dirt, and other substances to the ball to make it harder to see and help it dart and dive away from batters. A rule partly banning the spitball—each team was allowed to designate two pitchers who could still use the pitch—went into effect at the beginning of the 1920 season. Indeed, Mays himself was one of the pitchers who was still allowed to throw the spitter. This change started baseball's power pendulum swinging. In 1920, facing fewer spitballs, Babe Ruth would shatter his own home run record of 29 by hitting 54 round-trippers.

But Chapman's death gave baseball an incentive for more profound change, through both legislation and enforcement. After Chapman's death, the spitball ban was strengthened.* And instead of using only a handful of baseballs for each game and even retrieving foul balls from the stands, umpires began supplying fresh, white balls every time the ball got dirty.

"Long before MLB made batting helmets mandatory," wrote Jonah Keri in *Grantland*, "it banned doctored pitches and made umpires replace dirty balls regularly during a game, doing more to alter the game than perhaps any other rule change of the past one hundred years."

The very same factors that made it easier for a player to get out

---

* As a compromise, 17 active spitball pitchers were grandfathered in, including future Hall of Famer Burleigh Grimes, who continued to throw doctored balls legally until 1934.

of the way of a dangerous pitch also made it much easier for hitters to see a meaty fastball and drive it. And those fresh baseballs traveled farther when hit than the game-softened ones used only a few years before. It took both sides of the equation—legislation and enforcement—for the Ray Chapman tragedy to end baseball's "dead-ball era" and begin several decades of hitting-happy baseball.

And so it was in the NFL in the late 1970s. Stingley's tragic injury encouraged the league to add a few additional safeguards, notably rules preventing hits to the head and protecting the "helpless" receivers. But most important, this dark day provided an impetus for officials to enforce these new rules more strictly.

"Safety was still our number one concern," Shula explains. "It's devastating to see something like this happen. You had to look at it and see if there was something you could do to keep it from happening again."

These changes created a new environment where football could evolve, making passing much easier and more rewarding. Bill Walsh, who had become coach of the San Francisco 49ers, would discover that the attack he had created for Virgil Carter was now perfect for this new-school football, and he built a very different kind of dynasty upon these ideas. All because of a tragedy and the changes it inspired.

CHAPTER 8

HOW IS A QUARTERBACK
LIKE YOUR LAPTOP?

Imagine that the San Francisco 49ers are driving toward the goal line in some long-ago Super Bowl in some alternate universe. They're down by 4 points. It's 4th and goal at the 8-yard line with ten seconds to go. It's the final play of the Big Game. Joe Montana drops back and finds All-Pro wide receiver Jerry Rice in a corner of the end zone.

And in that brief moment, between when Montana releases the ball and it gets to Rice, all the lights in the stadium go out. At an even more inopportune time than in Super Bowl XXXV. For a long, strange moment, until the emergency generators kick in, the stadium is pitch dark.

And fans must simultaneously consider the only two mutually exclusive possibilities:

The pass was complete, and San Francisco wins the game.

Or it was dropped, and the 49ers lose.

This scenario may seem strange to us. But not so much for Montana and Rice. Like all modern NFL passing tandems, they work in the world of probabilities. Physicist Erwin Schrödinger would get it too. To explore the concept of probabilities in quantum physics, the Austrian physicist created a similar scenario, which included a cat, a box, and a vial of cyanide.

Schrödinger's Cat is a thought experiment. It's different from a real experiment in that it's something that can be pondered but not actually attempted in a lab. Another example of a thought experiment might be: "What would happen to the earth if the sun disappeared?" You can mull the possibilities and run the calculations, but of course you can't actually make the sun disappear. So it is with Schrödinger's Cat, albeit for slightly different reasons.

Imagine a cat. And imagine a metal box large enough for the cat to fit in. Then add a Geiger counter that's attached to a small hammer, which is triggered when the Geiger counter detects radiation. Also add a small amount of a mildly radioactive substance that has a 50 percent probability of emitting a single measurable particle over the course of an hour. And finally, a small glass vial of cyanide.

All of these things—cat, Geiger counter, cyanide, and radioactive material—are then placed into the box. If the radioactive material gives off a particle, the Geiger counter detects it, and the hammer breaks the glass and releases the cyanide.

If not, the hammer remains still, and the bottle of cyanide remains intact.

The power of the analogy is that it takes the mysterious things that happen at an atomic level and relates them to something as commonplace as a cat.

The box with the cat, Geiger counter, cyanide, and radioactive material is closed and a timer is set for an hour.

At the end of the hour, the timer buzzes.

What is the state of the cat?

Until you open the box, of course, you don't know. You know the possibilities: The cat is alive. Or the cat is dead. In the real world, it's an either/or situation. The cat can't be somewhere in-between, 50 percent alive and 50 percent dead. But a quantum physicist will tell you that the best way to think of the cat is exactly that: both alive and dead at the same time.

What was Schrödinger trying to get across? He was poking holes in some of the seemingly outlandish concepts that quantum physicists were postulating at the time. Physicists were arguing that an electron, which normally spins clockwise or counterclockwise, could, under certain conditions—called superposition—actually be rotating in two different directions at once. Schrödinger's Cat is designed to illustrate how ludicrous this is. This elegant, vivid analogy—which is said to have fascinated Albert Einstein—made Schrödinger's point so forcefully that scientists and laypeople are still mulling it over almost a century later.

There's one little problem: Laboratory experiments have since proven that at the quantum level matter behaves differently, drastically so, and that particles can indeed hold two seemingly opposite and mutually exclusive states at the same time. The quantum cat can be both dead and alive.

The West Coast offense doesn't function on a truly quantum basis. But it does prompt a way of thinking about football that's focused on probability.

The plays of the Lombardi era unfolded in a predictable, Newtonian way. On running plays, the player and the ball moved toward the goal. On passing plays, the wide receiver tried to get open, and the quarterback would throw the ball to him. The players were impressive athletes, to be sure, and the execution was impeccable, but conceptually the plays weren't that different from the ones you might run in a pickup game on the front lawn on Thanksgiving morning.

But as the gap between the best players and the worst narrowed, teams needed to find advantages that were more strategic than tactical. At the same time, other forces, ranging from concern about player safety to a desire to make football more exciting, drove a series of rule changes designed to open up the game. And thus was born the West Coast offense.

Ken Anderson was a math major. And that, as much as his size or his strong arm, may be the reason why he was the first quarterback to run the West Coast offense straight out of college. After his career at Division III Augustana College, Anderson was drafted by Bill Walsh for the Cincinnati Bengals in the third round of the 1971 NFL draft. "I think he liked my throwing motion," Anderson recalls. "I like to think I had a certain amount of intelligence he liked. I could process information pretty quickly."

As we noted earlier, Walsh devised the West Coast offense when Virgil Carter, who possessed a weak arm but a strong, nimble mind, became his starting quarterback, and this unconventional offense placed the quarterback in an entirely new role. The West Coast offense is built on a fundamental asymmetry that lies at the very heart of the game: The offense starts the play, and the defense has to react to it. Traditionally, there was a fundamental problem with increasing the complexity of a play. The very same complexity designed to confuse the defense also confuses the offense, disrupting their execution.

It's all about timing. Even a great offensive line gives a quarterback only three or four seconds to throw the ball before the pass rushers will knock him to the turf. This very real time limit constrained a traditional offense's ability to run complex plays. But Walsh's revolutionary offense found a way around that. The system increased the complexity for the defense while at the same time simplified the process for the quarterback and the receivers.

Here's a quarterback's-eye view of an old-school passing offense.

"When I was in Chicago, we'd look at film of a pass play, and there would be five receivers out in a pattern," explains Virgil Carter, recalling his tenure with the Chicago Bears. "We'd look and see where they would go and then the coach used to say, 'Well that was a pretty good pass play. Let's put it in our offense.' " Carter explains that the Bears would try and run the same play, but with no real understanding of why the receivers were running a particular route or what their alternatives were if the defense changed formation. "We'd send five guys out," Carter recalls, "and the quarterback's job was to find the open guy."

Even then Carter saw the fundamental problem with that approach. "You can't do that very well patting the ball with your left hand looking for an open guy," he explains. "In the NFL, you've got to get rid of the ball."

Contrast that with Ken Anderson's explanation of Walsh's West Coast offense.

"Bill's offense was a progression offense where you had a first, second, third, and fourth receiver," Anderson explains. "You have a primary receiver that you're going to go to first. A secondary, if that guy's covered. And there's a third receiver that's an alternative. And usually some kind of flare control to a running back or tight end that's a fourth in your progression. The whole theory is the defense can't take everything away. If you can get through the progression quickly enough before the pass rush gets to you, you can find someone that's open for a completion."

There's an elegant simplicity to Walsh's offense, so much so that it can actually be shoehorned into a rhyme. Sam Wyche, who was Carter's backup* during the early days of the West Coast offense, turned his progression into a little couplet: "1–2–3 Guarantee."

Listen to Anderson breaking down an actual play, and the inherent advantages of the West Coast offense become clear. The quarterback could memorize the simple steps of the progression and execute them under immense time pressure.

"The first play I learned was Split Right 82 Z In. The fullback is going to free release, run a wide flare to the strong side. The flanker

---

* Wyche had also been Greg Cook's backup and replaced him in the game in which Cook hurt his shoulder.

and the tight end are going to do 12-yard curls. If [a defender] jumps on the running back, you throw to the flanker. If he stays back, you throw to the fullback. And if they're both covered, you throw to the tight end. And finally you'll get to the backside progression, which would probably be a backside flare or check down to the halfback. It's one to the fullback, two to the flanker, three to the tight end, four for halfback. If the strong side linebacker comes [on a blitz], the quarterback's got to throw hot, and the flare to the halfback is the hot throw."

Notice the very clear set of priorities, like a professional organizer's to-do list. Instead of a single overly general objective—"find the open guy for a completion"—the quarterback in Walsh's offense is instead presented with a series of specific tasks in a precise sequence—"One to the fullback, two to the flanker . . ." Under any circumstances, this kind of detailed task list is an effective productivity tool. When a 290-pound defensive end is about to knock you down and grab the football, this level of specificity spells the difference between success and failure.

Walsh's progression is not only about compressing a series of decisions into the shortest amount of time. It's about linking those decisions to the actions of the defense. Carter explains that each step of the progression led to a yes-or-no decision, choreographed to each step of his drop, depending on the coverage he saw in the defense.

"On every single pass that Bill Walsh designed, the quarterback, within the first step or two, had a key," Carter explains. "If it's a four-three defense and the middle linebacker retreated straight back or to your left, you knew that you were going to go to the right. You could forget the left side of the field. On step three and four, you would look to the right. If the strong safety was taking the fullback and the flanker or covering the tight end on an inside release, then you knew the wideout was your target on an out or a curl. When you hit your fifth step you were delivering the ball to a guy who pretty well was open. And if there was any other change in that pattern, the worst case was the back in the flat was wide open. You got him the ball, then he could run for yardage."

How is a quarterback's progression like a spreadsheet on your laptop? They're both powered by binary language. That's when a complex problem is reduced to a series of simpler questions that offer only two mutually exclusive options. The Quarterback's Cosmic Checklist—Yes, the safety is blitzing / No, he is not—runs on this simple but very powerful mode of communication. So does Morse code. So does everything from a smartphone to a nuclear power plant.

In a computer, the two positions of a simple switch—voltage on and voltage off—become a language that can be understood by a computer's central processing unit. A series of 1s and 0s in a specific order can be used to represent an alphabet, and more important, they can encode all the rules of math in a very compact algorithm. A simple calculation written in decimals might take as many as 100 rules, while the same calculation in binary can be accomplished in only four.

This brand of mathematics is called Boolean algebra, named after George Boole, a nineteenth-century mathematician and philosophe. His use of 1s and 0s and true-or-false statements is the computer equivalent of the yes-or-no statements found in human languages—or Bill Walsh's quarterback reads.

For more than half a century after Boole's death, Boolean algebra was mostly of interest to mathematicians, but in 1937 a clever graduate student from MIT, Claude Shannon, recognized that Boolean logic could allow electronic circuits to "speak" with each other. Simple switches communicated with each other by turning on or off, which can be read as a one or a zero. If the switches are arranged along Boolean logic, then the two simple digits can represent large numbers and complex mathematical operations. In 1937, Shannon laid the foundation for the circuits that Bell Labs used to route telephone calls to their proper destinations.

These yes-or-no propositions also became the foundation of computer programming. A hard drive uses the north and south poles of a magnet to store digital data. The spinning disc is covered with a thin magnetic film and a read-write head hovering above it which "senses" the direction of the magnetic regions (or bits) on the disc. CD, DVD, and Blu-ray players work in essentially the same way, except that small pits of long and short lengths are read by a laser replacing the magnetic head of the hard drive. Shannon's binary language provided the seed for a new field called information theory, which now explains everything from cryptography to gambling probabilities.

In essence, Bill Walsh was trying to get his quarterbacks to think like a computer. He told them to forget about the shades of gray that quarterbacks of previous generations like Johnny Unitas or Terry Bradshaw applied to "finding the open man." In a traditional offense, a quarterback had to scan the field to determine the defensive coverage and then find that open man. At the best of times, this was like doing a Where's Waldo puzzle; at the worst of times, it was like answering an essay question on an exam for a class for which you hadn't done any of the reading: "Compare and contrast the current defensive coverage with the pass routes assigned to the play in question. Please be specific." When this kind of impressionistic task was accompanied by pressure from the pass rush, the result was unpredictable and potentially disastrous. The quarterback was taking an analog view of the whole field in much the same way that a phonograph stylus reads the whole frequency spectrum encoded on an LP record.

In his West Coast offense, Walsh got his quarterbacks to see the world in black and white. He took a complex situation and broke it down into simple yes-or-no decisions of the kind that microprocessors can make with lightning speed. He then built these true-or-false tests into the kind of if-then decision trees that are the center of computer programming languages. In short, Bill Walsh made football digital.

When he joined the 49ers, Jerry Rice had a distinctive habit. Whenever he caught a ball, the young wide receiver wouldn't stop. Even in passing drills, where there was no defender to beat, Rice would catch the ball and just keep on running until he reached the goal line.

"I wanted to make the catch, and then I would sprint sixty to eighty yards," Rice recalls. "It becomes like a reaction. I wanted to catch the ball and score."

Rice looked at the reception in a fundamentally different way from any receiver before him. For Rice, the catch wasn't the *end* of the play, it was the *beginning*.

Rice might just be the greatest football player of all time. The argument for his supremacy goes something like this: Rice played a key role on one of the greatest teams in football history. He also collected a remarkable number of single-season and career records. Rice dominates the record book for receivers, from the most touchdowns in a season to the most receiving yards in a career. But the most important record that Rice holds is an unofficial one: yards after the catch. For most of Rice's career, the NFL did not officially keep track of that stat.

But it's this statistic that lies at the heart of Rice's third and most important claim to greatness: He changed the game. Before Rice came along, catching the ball was an end in itself. Rice made it a *means* to an end. The West Coast offense turned the pass from a risky, hit-or-miss strategy into a long handoff. When the San Francisco offense was functioning at its best and Joe Montana had enough time to work through his progression, Rice would catch the ball without breaking stride, his role morphing seamlessly from that of a receiver to that of a running back.

"I had football speed. If I ever got out in front of you, you couldn't chase me down," Rice explains. "I used fear to motivate me. They're really trying to hurt me, but they can't hurt me if they can't catch me."

And while other players, notably San Francisco running back

Roger Craig, had already begun exploring this key advantage of the West Coast offense, it was Rice who brought those experiments to fruition.

"The West Coast offense is all about the timing," says Rice. And then he amends his explanation. "It's *anticipation*." With apologies to Carly Simon, in Rice's version of the offense, anticipation is all about *not* keeping him waiting. The quarterback has to have the confidence to throw the ball to, well, *nothing*. To a spot in thin air, anticipating that by the time the ball arrives at that spot, Rice will be arriving at the same time from a different direction.

"If Joe has a five-step drop, the ball is going to be in the air before I come out of my cut," Rice explains, "If he tries to throw it *at* me, it's not going to time out."

This improbable precision is football's answer to the flying trapeze. One acrobat leaps into thin air, and just as it seems like he's about to plummet to earth, his partner swings over in the nick of time. Under the big top, fans inch toward the edges of their seats, holding their breath, and explode in applause as the daring young man defies death once more. In the West Coast offense, this split-second precision is just 2nd down.

The idea that timing is everything actually predates the West Coast offense. It can be traced back to one of Walsh's mentors, that hugely influential but somewhat forgotten coach named Sid Gillman. At the same time that Vince Lombardi was bringing the scientific method to the running game in the National Football League, Gillman, who had also been an assistant coach at West Point under Red Blaik, was doing roughly the same thing in the rival American Football League for the passing game. Gillman isn't a household name probably because he never won a Super Bowl and because he coached most of his career in the now-defunct AFL. He won one AFL Championship, in 1963, before there was a Super Bowl. Gillman's Chargers won five AFL West division titles between 1960 and 1965, but by the time the Super Bowl came around in January 1967, the Chargers had lost ground in the division to the Oakland Raiders and the Kansas City Chiefs, and the Houston Oilers team he coached in the NFL in the early 1970s wasn't a championship contender.

But a look at Gillman's "coaching tree" illustrates his influence. Those coaches who served as assistants under him or worked under Gillman's own assistants read like a Who's Who of modern NFL coaching: Chuck Noll, John Madden, Joe Gibbs, Dick Vermeil, Tom Flores, Mike Holmgren, Tony Dungy, George Seifert, Mike Shanahan, Jon Gruden, Mike McCarthy, Brian Billick, Mike Tomlin, and John Harbaugh. Among them, Gillman's disciples won a total of twenty-five Super Bowls. That group also includes Bill Walsh.

"Being part of Sid's organization was like going to a laboratory for the highly developed science of professional football," said Raiders owner Al Davis.

Gillman brought a scientific rigor to professional football. He was among the first coaches to study film. He would assemble a season's worth of plays and then group them by category: off-tackle plays, screen passes, fly patterns. Contemporary coaches, like Jon Gruden, who watched game video until the wee hours of the night and fell asleep on their office couches had Gillman to thank for it.

But mostly what Sid Gillman brought to pro football was a passing game executed with scientific precision. Because of a variety of rule differences, the passing game was much more effective in the AFL than it was in the NFL, and Gillman's passing attack ranked among the best in the league. In 1964, Gillman was frustrated by the persistent sloppiness in the execution of the Chargers' pass patterns, a problem that wouldn't yield to increase the amount of practice. So he sent his receiver coach Tom Bass back to San Diego State University to talk to the math professors there about how to sync up the team's future Hall of Fame receiver Lance Alworth with quarterbacks John Hadl and Tobin Rote.

The answer? Trigonometry.

Bass and the professors quickly figured out that pass patterns could be best explained in terms of triangles. The mathematicians treated the gridiron like a big piece of graph paper. One leg of the triangle would be the receiver's downfield progression, the distance from the line of scrimmage to the point where he'd make his break. The second leg is from that breaking point to the point of the intended catch. The third and longest leg of the triangle is the distance from the quarterback to the point of the catch. Like a trigonometry word problem, the goal was for Alworth to cover the first two legs in the same time it took for the quarterback to drop back and throw the ball along the long leg of the triangle.

It depended on where the ball was placed in relation to the hash marks on the field, the professors told Bass. Alworth and the other receivers would have to line up a different distance from the sideline on each play in order for the timing to sync perfectly.

"We didn't talk about the hypotenuse," Bass laughs.

Bass's geometry took Gillman's offense to another level, although it remained a high-risk, high-reward attack focused mostly on the long pass, the same kind of offense that Walsh used with strong-armed Greg Cook during his early seasons in Cincinnati. Walsh's West Coast offense would take precision timing to the next level.

One fundamental problem with a traditional running play was its predictability. It starts within a few feet of the snap of the ball, and even the most brilliant runners—from the powerful Jim Brown to the elusive Barry Sanders—are limited by this. But what if the ball could be handed off *anywhere* on the field?

This was the promise of the West Coast offense.

The payoff was huge, but the cost was precision.

"You've got to have players who are intelligent and are willing to execute. You put the time in during the week," Rice explains. "You build timing. You know exactly what the quarterback's thinking. He's on the same page."

"Bill's offense, in the early days—everything was based on precision," adds Anderson. "All of his quarterbacks were so drilled in their footwork, so that your number one receiver was coming open as your plant foot hit, which allowed you to get through a progression in a two-and-a-half- to three-second time period before the rush. The depth of the route had to time up with the drop of the quarterback. Everything had to be very precise. The secondary route, by the time the quarterback is getting to him, is coming open just at that point in the progression. We spent a lot of time and a lot of thought on that. Which is why all the quarterbacks worked so hard on the mechanics of their drop so that the depth of the route would time up with the drop of the quarterback."

If he was to be there when the ball arrived, Rice needed to be every bit as precise as the quarterback.

"It's almost like a dance," Rice explains. "As a receiver, if I've got a 12-yard out, where I go up 12 yards and I break to the sideline, I've got to count in my head six steps. Once you've got everything on the same plane, where it's the precise timing, it's hard for a defender to knock the ball down and make an interception."

In the same way that a quarterback's progression is synched to the defense's coverage, Rice's patterns depend on the tactics a cornerback is using to attempt to stop him. "Defenses change on the snap of the ball," he explains. "We have certain adjustments we have to make on the run. You've got to be like a quarterback as you're running down the field reading the defense."

As an example, Rice uses the play 200 B Slant, which was designed with him as the primary receiver. While the play as written is supposed to be a simple slant pattern ending with a pass from Joe Montana, his actual route depends on how aggressively the defender is playing him. "The defensive back, if he's off, [playing me] back five or six yards and not up in my face, then I'm going to run a three-step slant. I don't have any options," Rice explains. "If he's up on bump and run, now I've got to be able to work him off the line of scrimmage and try to get inside. If I can't get inside, then I make an adjustment and try to make it into a fly route, where I'm just trying to beat him and Joe can release the football," he says. "You don't want to become a robot out there. You've got to be a football player."

Rice's goal is simple, even if the execution isn't. Arrive when the ball does. No one in the history of football did it better or more often.

Without a blackout or a cat in a box, the West Coast offense proved its value at the very beginning of Bill Walsh's run. With a trip to the Super Bowl on the line in the NFC Championship Game and time running out, Joe Montana's 49ers were driving toward the Dallas goal line.

When the Dallas defense flushed Montana out of the pocket to the right, most of America expected that it was a broken play and he'd just toss the ball out of bounds. Carter, a veteran of the West Coast offense, anticipated something else.

"That's a play that we would practice. I would move right, the wideout would go down the field and clear, the back would clear in the flat, and the opposite wide receiver would come across the field about fifteen yards deep. If I rolled right, that's what I would expect to find. The receivers all had contingencies, and that's exactly what happened in The Catch," Carter recalls. On his TV screen, Carter watched Montana fling the ball in seeming desperation, but he wasn't at all surprised when Dwight Clark came out of nowhere to snatch it before it sailed out of the end zone.

"Dwight Clark was the opposite wide receiver coming across ten, fifteen yards deep. Montana knew he was going to be there. It was something that they had actually designed, and that's the way all the plays were in the West Coast offense," says Carter. "Of course it was a great catch and a great throw. But it wasn't random chance."

CHAPTER 9

SAM WYCHE AT PLAY IN
THE FIELDS OF CHAOS

Sam Wyche knew he was onto something when he saw the way Renaldo Nehemiah was breathing.

"Jeepers, Skeets, I thought you were some kind of world-class athlete? Here you are breathing like a racehorse," Wyche said to Nehemiah, using the wide receiver's nickname. Wyche was the passing director of the San Francisco 49ers, and Nehemiah was not only one of the fastest players in the NFL, he was also an elite track athlete, a former Olympian, and a multiple world-record holder in the 110-meter hurdles.

Nehemiah turned around and looked Wyche straight in the eye. "Sam, the reason I'm breathing heavy is I just ran 60 yards as fast as I could run it." Nehemiah took one more breath and added, "If you give me about four more seconds, I will be breathing through my nose again."

"It immediately clicked," Wyche recalls. "If I could get practice at a tempo where my team is recovering in eighteen to twenty seconds, and my opponent has been practicing all week at about a minute a play, and I snap the ball at about twenty-two to twenty-three seconds, I am playing against a slightly fatigued team on play number two. On play number three, they're slightly more fatigued. And pretty soon chronic fatigue sets in and they start tapping the side of their helmets saying, 'Get me out of here. I need a break.' You are now playing against a lesser opponent. You are gaining an edge."

Wyche understood the importance of that encounter with Nehemiah and the epiphany that stemmed from it. But he didn't act on this revelation, at least not right away. Wyche wasn't coaching the 49ers; Bill Walsh was. San Francisco had already established a dynasty based on Walsh's state-of-the-art West Coast offense. So Sam Wyche filed that moment away.

But in the fullness of time, this casual encounter between a coach and a player on the practice field on an otherwise normal afternoon would be the act that would introduce chaos theory to professional football.

The huddle may have been born in the last years of the nineteenth century, when a deaf quarterback named Paul Hubbard called his Gallaudet University teammates around him to keep the opponent from seeing the sign language he used to call his plays. Or maybe it was Penn that began the practice in 1894, because All-American center Alfred E. Bull was hard of hearing and couldn't catch the plays at the line of scrimmage. Or maybe it was Amos Alonzo Stagg's University of Chicago Maroons, who had trouble hearing the play calls against Michigan in 1896. Or it may have been started by Borden Burr, a University of Alabama quarterback who took credit for the innovation, claiming that he had been "knocked dizzy" during a game in 1895 and couldn't remember the signals, so he gathered his teammates around to talk it over. Or the huddle may have really originated in 1918 when Oregon State head coach Homer Woodson "Bill" Hargiss, frustrated by the way that rival University of Washington seemed to anticipate every play, instructed his players to gather 10 yards behind the ball and whisper the play to each other. Then again, Lafayette coach Herb McCracken may or may not have been the huddle's originator when he noticed that Penn had scouted the team's previous games and instructed his players to gather behind the line of scrimmage and call the play in secret. Or wait—the huddle might have actually debuted in 1921 in Illinois under coach Bob Zuppke, who was the subject of criticism for slowing the game down, but silenced his critics by winning a national championship with it in 1923.

Then again, the origin of the huddle isn't that important. What matters is that the huddle took a while to catch on. But once it did, it became an integral part of the game.

And no matter when it was born, the huddle was part of a trend that can be traced back to Yale coach Walter Camp's innovation of game planning and play calling at the end of the nineteenth century. In its earliest days, football was based on free-flowing action with few breaks, much like modern soccer or rugby. Camp's rule changes broke the game up into discrete "plays," which borrowed from baseball's more deliberate, episodic pacing. The plays may have lost something in terms of continuous action, but they gained something too. A playwright would understand that the plays were similar to scenes in a play, each with a conflict (*To pass or not to pass, that is the question*) and a resolution (*Incomplete. 3rd and 7*).

But a scientist would understand that the extra time between plays provided the opportunity to analyze the effectiveness of each effort, echoing the way that a researcher, or an inventor like Thomas Edison, runs a trial and then parses the results. Improvisation gave way to deliberation.

The huddle worked at another level as well. It established clear communication among the players on the offense so that they could execute their plays with precision. The huddle sacrificed the element of surprise in favor of predictability.

Until Sam Wyche came along.

———

A couple of years after watching one of the world's fittest athletes huffing and puffing between plays, Wyche became the head coach of his own team. The 1984 Cincinnati Bengals were very much the opposite of the 49ers. They were on a downhill slide since meeting San Francisco in the Super Bowl three years earlier, and their offense was particularly problematic.

So one day during training camp at Wilmington College, Wyche was reminded of that moment with Nehemiah and decided to try something. In the middle of an otherwise typical practice, Wyche called a "nickel period," where the Bengals offense would run a series of long 3rd-down plays—from 3rd and 6 up to 3rd and 10—while the team's defensive coordinator Dick LeBeau, a Hall of Fame player who would later become the defensive coordinator for a Super Bowl–winning Pittsburgh Steelers squad, practiced his nickel defense, which swaps out a linebacker for a fifth defensive back.

"If it is a 3rd down and 8, everyone in the stadium knows you are going to pass," Wyche explains. "So the defense brings in their best upfield pass rusher instead of the slow, run-stopping nose tackle. They take out the heavy-legged middle linebackers and put in another defensive back or two, and now they have got all their good cover people in and their pass rushers in. They know we are going to pass. And sure enough, we do. And sure enough, they win a percentage of the time. I said, 'This is stupid.' "

Wyche realized that one small thing ceded a huge tactical advantage to the defense: the huddle. So Wyche then called a "move the ball" period, in which the offense and defense would go at each other under game conditions.

"We'd simulate another 1st and 10, then a 2nd and 8, then a 3rd and 8," Wyche recalls. "I told my offensive guys, 'When we get to the third down, we are not going to huddle up. We are going to run this play. You offensive guys understand that? And don't say anything to your buddies on defense.' So we did that. Dick starts to substitute, and we would run up to the line of scrimmage, snap the ball, and throw it down the field to an uncovered guy. Dick is screaming and hollering, 'You can't do that!'

"'Let me explain something to you, Dick. You *can* do that.'"

And all afternoon, the Cincinnati offensive players would rush back to the line of scrimmage like they had a plane to catch, then take advantage of the mismatches that were created when LeBeau's defense had to scramble to make substitutions on the fly.

What started out as a way to shake up a dull training-camp practice—and to provoke a reaction from Wyche's uptight defensive coordinator—evolved into a full-blown strategy.

The no-huddle—or the hurry-up, as it was sometimes called—wasn't invented by Wyche. It was a tool that was used by every team when trailing at the end of a close game. But it was a tactic of last resort, and other coaches would never consider using it all the time, any more than they'd start a game with an onside kick.

Wyche reinvented the huddle, and instead of circling up 8 yards behind the line of scrimmage, his offense would confer right by the ball. "We called it a sugar huddle," he explains. "I said 'Guys, I want you to get close to that line of scrimmage, kind of sugar up like you would to your sweetheart.' So the center only has to take one step, and he's at the line of scrimmage. The quarterback just has to turn around, and he is standing on the center."

And on most plays Wyche went even further and tossed out the huddle altogether, which yielded more than just a chance to keep the nickel defense off the field. "Most teams work at about one play per minute in practice. Well, in the no-huddle we wanted to get five plays every two minutes," Wyche explains. "That would be our pace and we could do it in a game."

As a journeyman quarterback fighting for his job—and sometimes his life—Wyche had learned to watch the defenders for small clues as they lined up across from him.

"The defense talks to you with their body language," Wyche explains. "If you see a defensive back bending at the waist, he is probably playing zone. If you see him bending at the knees, crouching straight down where he has got to chase a guy all the way across the field? That means he is playing man to man."

Wyche knew he was onto something when he saw how his own defensive backs were standing as the offense threw play after play at them.

It's all in the thumbs. When a player is feeling good, rested, and ready for the next play, Wyche explains, he stands up straight in a posture with more than a little Superman-style swagger: hands on hips, fingers across his belly, thumbs pointing back. But when a player is tired, his posture changes dramatically. He hunches over, looking toward the ground, fingers down almost on his butt. And his thumbs? They're pointing *forward*.

"It tells the offense they are wearing down," Wyche says. "And as soon as you see it, speed it up, cut another couple of seconds out."

When Wyche saw his first-string defenders sucking wind and his world-class defensive coordinator struggling to keep the right players on the field, he knew that he had stumbled onto something. His no-huddle offense was an innovation that would take his Bengals all the way to the Super Bowl and leave them one miraculous Joe Montana drive short of a championship.

What exactly did Wyche stumble onto? In a word, chaos.

What exactly is chaos? We all instinctively understand chaos, at least as it relates to a junk drawer or a seven-year-old's birthday party. But in science and mathematics, chaos has a more precise definition.

Here's how Edward Lorenz, the creator of the Butterfly Effect, defined chaos: *When the present determines the future, but the approximate present does not approximately determine the future.*

To put it another way, chaos is about how small things can have big effects. And few people in the world understand this more deeply and with more nuance than Stephen Wolfram.

Described as "the Bob Dylan of physics," Wolfram was a member of the first class of MacArthur Fellows, winning the so-called Genius Grant in 1981 at the age of twenty-one, along with heavy hitters like Stephen Jay Gould, Derek Walcott, Henry Louis Gates Jr., and Robert Penn Warren.

And the Dylan comparison is an apt one. Wolfram sports an oversized personality, and trying to pigeonhole him can be an exercise in futility. Wolfram was doing cutting-edge work in particle physics as a teenager, then suddenly shifted his attention to the more practical—and populist—work of developing a computer algebra system called Mathematica. He spent a decade working on a book called *A New Kind of Science* which tackles the biggest questions in science. He argues that an understanding of simple computational systems can lead to a new and deeper understanding of biology, chemistry, and physics. On the other hand, he admits, somewhat sheepishly, that he's collected more data on himself—tracking everything from keystrokes to heartbeats—than any other human being has.

The English scientist's ideas are compelling because they're at once mind-bogglingly complex and captivatingly simple. "Wolfram goes on to explain that by applying a single key observation—that the most complicated behavior imaginable arises from very simple rules—one can view and understand the universe with previously unattainable clarity and insight," wrote Steven Levy in a *Wired* magazine profile of Wolfram. "The idea of complexity arising from simple rules—and that the universe can best be understood by way of the computation it requires to grind out results from those rules—is at the center of the book. The big idea is that the algorithm is mightier than the equation."

Wolfram admits that he knows almost nothing about American football—and apologizes for it—but he does understand complex systems and is happy to share his insights.

"The knob of chaos theory," he says, "is the dependence on initial conditions."

In explaining the hidden power that can be unleashed when you make even the smallest changes in those all-important initial conditions, Wolfram comes up with a delightfully outrageous analogy that would fascinate both a small child and a physics postdoc.

"Take flipping a coin. The reason that flipping a coin leads to what seems like random outcomes is a little subtle," says Wolfram. "If you had a machine flipping the coin and the machine was nicely calibrated, it flips the coin, the coin spins around fifteen times or something, and it *always lands the same way up*. The laws of motion for the coin are always the same, and if you flip the coin the same way, it will always land exactly the same way up. The reason why it *seems* kind of random is that when we flip the coin, our brains and muscles aren't that precise. We *think* we're flipping it the same way, but if we flip it just a tiny bit harder, it'll turn one more time in the air and come up on a different side. That uncertainty in our initial condition, which we as humans generate, makes it *seem* like a random process."

It's a lot like Edward Lorenz's Butterfly Effect, where the tiniest disturbance in the atmosphere can cause a tornado halfway around the world.[*]

It's also kind of like Wyche's no-huddle offense.

With Wyche's groundbreaking offense, there was no actual change in the plays the Bengals' offense was running. They were still using the exact same pass plays and running plays in a similar ratio. By shortening or eliminating the time between plays, Wyche was changing the framework within which those plays were run. Or, as Wolfram would say, the *initial conditions* of the play.

And, as Wolfram would have predicted, that small change had profound and unexpected consequences. What had once been a given—that there would be a predictable gap between plays—was suddenly up for grabs.

---

[*] Wolfram, though, is quick to point out that the Butterfly Effect is more useful as a metaphor than a practical example. In the real world, the flap of the butterfly's wing would be damped out by the viscosity in the atmosphere; only in the rarest of circumstances would a factor that small actually change the weather.

Wyche tinkered with every aspect of the huddle. He would sometimes bring his entire team—not just the quarterback—to the sidelines to discuss the upcoming play. It looked a lot like a time-out in
basketball, with the advantage being that every player would now
hear about the play directly from Wyche or his coordinators. It also
had a few subsidiary benefits, allowing players to get some quick
medical treatment, like bandaging a cut or fixing a minor equipment
malfunction like a broken chinstrap.

At other times, the Bengals would put twelve or thirteen men on
the field when the ball was being spotted, and then get down to the
required eleven by having the extra player or players run off the field
just before the snap. "When there's no huddle, it's okay to have more
than eleven players on the field as long as they're off before the ball
is snapped or before the clock runs down," Tony Veteri, the NFL's
assistant supervisor of officials, told *The New York Times* at the
time. Since the offense controls the timing of the snap, there was
little danger of the Bengals getting caught with too many men on
the field. But this tweak had defensive coordinators pulling their hair
out; the uncertainty prevented them from making substitutions, because they risked a penalty for having too many men on the field
when the Bengals snapped the ball.

"It became an offensive strategy, just like sending a man in motion or playing two tight ends or playing four wide receivers or taking the quarterback out from center and going to the shotgun,"
Wyche recalls.

Like so many innovations, the no-huddle wasn't an immediate
success. In its earliest iterations, the very changes designed to keep
the defense off balance instead kept the offense off balance. And the
beginning of the 1987 season was disrupted by a players' strike, making it exactly the wrong time to introduce an intricate new offense.

But eventually the Bengals stopped outsmarting themselves.
Young Boomer Esiason replaced Ken Anderson as the Bengals'
quarterback and he embraced the system wholeheartedly. The no-
huddle required him to memorize plays by the dozen, with nothing

more to help him in a game situation than a quick glance over to Wyche or his assistants.

Esiason could throw to a collection of receivers who were flexible enough to work within the system's framework. "We had kind of the perfect storm," Wyche explains. "We had a tight end named Rodney Holman, who was as good a receiver as most wide receivers. So we did not have to take him out when we wanted to throw the ball. We had a running back named James Brooks from Auburn, who was a good receiver down the field, so we did not have to take him out and put in an extra wide receiver. So we could break the huddle after second down, we could go right to the line on the third down and be in a spread formation—which is four wide receivers, two on each side—with our tight end being spread on one side and our running back being part of the spread on the other side. It gave the defenses a lot of problems because they had to go out and cover people with linebackers, and they could not run with them."

In this way, the no-huddle represented a profound reimagining of the game. It was forward-looking, but in many ways it also represented a nod to the past. The nonstop action of the no-huddle evoked the continuous action of football's earliest days. And by depending on a cache of versatile players to play a variety of roles, the no-huddle was a throwback to the days before specialization took over the game.

It worked. It gave the Bengals short-term benefits in terms of personnel mismatches—a fleet wide receiver on a lumbering linebacker or a burly tight end covered by a tiny cornerback. But unlike many game-planning advantages that depend solely on the element of surprise—and hence tend to lose their edge the more they are used—the no-huddle actually worked *better* when the Bengals needed it most. As the Bengals moved toward the opponent's red zone at the end of a drive, the defense was sucking wind more desperately than they had at the beginning of Cincinnati's possession. By the fourth quarter, the opponent's legs were even more leaden—and defenders more apt to give up the big play—than they were in the first. The no-huddle strategy proved more effective late in the season, as Cincinnati faced teams depleted by injury and players battling long-term fatigue.

"We actually know what we're doing," Wyche said to skeptics at the time. "Sometimes we use it to keep personnel off the field, sometimes to change the tempo of the game, and sometimes to force defenses out of a certain coverage."

In 1988, the Bengals started the season 6–0 on their way to a 12–4 regular season, the best record in football.

During the regular season, the no-huddle had offered the team an early-adopter advantage—it was so far outside the norm that it didn't make sense for teams to develop an elaborate game plan to stop the no-huddle, because Cincinnati was the only team using it. Which is why many opponents, especially those outside the division, treated the Bengals like outliers and didn't devote many resources to stopping their unusual offense.

But as the playoffs approached, the stakes were raised. Cincinnati's opponents needed to find a way to counter the no-huddle.

Even if it meant bending the rules.

That's what the Seattle Seahawks did when they played the Bengals in the second round of the 1988 playoffs. The strategy centered on run-stopping nose tackle Joe Nash, who almost always left the game on passing downs.

"In that game," Wyche recalls, "Joe Nash would look to the sideline, and there was a coach that literally would make the shape of a gun with his pointer finger and his thumb up. If he pulled the trigger with his finger, the nose tackle who had been looking perfectly good would act like he had gotten shot and grab his knee or his hip and he would fall down. The referee would call time-out for an injury on the field. Of course Seattle would bring a nickel in. That is how they were going to stop the no-huddle, and they did it pretty much the whole ball game."

Wyche approached the officials, but because it involved an injury, they were powerless to act.

"I know what we're looking at," the official told Wyche, "but we are not doctors. We can't make that medical judgment."

"Keep your eyes open," Wyche replied, "because you can earn your M.D. right here."

Ultimately, the strategy didn't work, because while Seattle managed to throw Esiason off his game, limiting him to just seven completions, Seattle's offense was anemic and Cincinnati jumped out to a 21–0 lead.

In the AFC Championship Game, Buffalo Bills coach Marv Levy saw the potential in Seattle's approach and threatened to do the same thing. Just two hours before game time, NFL Commissioner Pete Rozelle sent an emissary to deliver a message to Wyche. The Bengals' coach, having been tipped off by the NBC broadcast crew, was ready for him.

"Well, gee, what are y'all doing in here, exactly one hour and fifty-five minutes before the game?" Wyche said for the benefit of the tape recorder he had running.

"Sam," said the emissary, "the commissioner told us to come tell you that you cannot run the no-huddle, because Marv Levy said he will fake injuries like Joe Nash did in the Seattle game, and this is too high profile a game for a mockery to be made of it."

"Well," Wyche countered, "that's like coming in to the fellow that owns a restaurant and saying, 'You know you could be robbed between six and eight o'clock at night, so you can't be open during those hours.'"

His joke delivered, Wyche got dead serious. "Go get the commissioner on the phone. I want to talk to him personally, because I want him to understand that if we lose this game and you enforce that rule, the first thing we talk about is the fact that the commissioner of the National Football League tried to affect the competitive balance of the AFC Championship Game. There are a lot of gamblers out there with a lot of firepower. I don't think I'd want that around my neck."

The NFL official left the room and—by Wyche's account—

returned in only twenty seconds. "The commissioner said go ahead and run the no huddle No problem," he said.

That's what Sam Wyche's Bengals did, riding their special brand of applied chaos theory all the way to the Super Bowl. As would Marv Levy. His Bills began running their own version of the no-huddle the following season. They would win three consecutive AFC Championships.

CHAPTER 10

PLAYING DEFENSE, HEISENBERG STYLE

A hacker can tell you that there are two basic ways to gain access to a secure computer system. You can build a strong, sophisticated computer that will monkey-brain its way through layers upon layers of encryption. The problem is that once you crack the code, the code writer can just up the ante by adding another layer of encryption. All of a sudden you're back to square one, but now you need an even more robust computer to hack through that new encryption.

The other way is to sneak into the office and find the password written on a steno pad in the secretary's top desk drawer.

Up against the game's increasingly complex offenses, football's defensive coordinators faced a similar choice. They could find guys who were bigger, faster, and smarter to provide a better man-to-man response to offenses that were increasing their complexity at a geometric rate.

Or they could change the way defense was played. Within the rules. And outside of them.

———

One by-the-book response to a modern passing offense is the zone blitz. This defense mirrors the West Coast offense in that it marries high- and low-risk strategies into a sophisticated hybrid that's designed to stop a West Coast passing attack. The zone blitz was popularized by Dick LeBeau, defensive coordinator of the Cincinnati Bengals, who would later go on to run the defense for two of Pittsburgh's Super Bowl–winning squads.

"It was on a cross-country flight following the team failure of 1987 where LeBeau drew up the Gutenberg bible of NFL defenses. With his tray-table down and a pen in hand, LeBeau began doodling on a napkin," wrote John Breech on the *Bleacher Report* website. "The doodles turned into safeties blitzing, but that had been done before. Then another doodle: defensive linemen dropping back into pass coverage to make up for the blitzing safeties' exposed area."

To understand the zone blitz, it's best to break it down into its component parts.

First the blitz. Even the name is edgy, conjuring up World War II fighter planes, and as Chris Brown observed on *Grantland,* "A blitz is the closest thing we have to football bedlam." In its simplest form, the blitz is one or more defenders abandoning pass coverage in order to increase pressure on the quarterback.

This makes the blitz a play of both great opportunity and great peril for a defensive coordinator. If the blitzing defender does his job, good things happen. At a minimum, he jars the quarterback out of his progression. Time is a quarterback's best friend, and losing a second or two can be his worst enemy. A harried quarterback makes bad decisions. Sometimes he throws inaccurate passes. Sometimes he scrambles when he ought to take the sack. And sometimes he throws into coverage when he ought to throw the ball away. And sometimes that ill-advised pass is intercepted.

But if the quarterback does pick up the blitz and doesn't panic, the balance of power shifts. A receiver will be wide open at the spot the blitzing defender just vacated. If the quarterback can get the pass off, the receiver can catch the ball unimpeded and run for a big gain. Former Florida State football coach Bobby Bowden once gave a lecture about the blitz with the apt title "Hang Loose—One of Us Is Fixin' to Score."

Zone coverage is exactly the opposite. It's a defense based on discipline.

Each player in the backfield is assigned not to a player but to an area of the field. The defender must display enough self-control to keep from chasing a receiver who ambles into his coverage area. He must watch the quarterback and only react when the play comes his way. Playing the zone challenges the defender's mind, because he has to wait for the play and then, once he sees that the ball is coming his way, commit fully to stopping the receiver.

LeBeau's zone blitz combines those two vastly different approaches in a way that eliminates many of the risks of the blitz while retaining its rewards. If a safety blitzes the quarterback, a linebacker moves into the area vacated by the safety. A defensive lineman then covers the area left open by the linebacker.

This last-minute shift creates huge problems for the quarterback. A modern passing offense begins with the "hot" read: Is the defense blitzing? When the quarterback sees the defense blitzing, he invokes his "guarantee" option, usually a short pass to the running back. But suddenly the fallback receiver, instead of being wide open, is now covered.

The zone blitz may look chaotic, yet when executed properly, it's anything but.

"The game is played on a rectangle, and within that rectangle, the offensive players fit into multiple levels that force opponents to defend the whole field," LeBeau told Ron Jaworski in his book *The Games That Changed the Game.* "Offenses were literally creating squares and triangles with their routes. I thought it might be a good idea to match those shapes with squares and triangles of my own. I wanted to put my people in areas where the offense was sending its players."

———

Werner Heisenberg was one of the fathers of quantum physics and won the Nobel Prize in 1932. His most enduring contribution—aside, perhaps, from being the namesake of a character on the television show *Breaking Bad*—is his Uncertainty Principle, which may be the quintessential example of quantum weirdness. The takeaway lesson of the Uncertainty Principle is that our understanding of the smallest particles boils down to trade-offs. You cannot know everything about an electron, and the more you know, the weirder it acts.

When we describe an object using Newtonian physics, we can determine its location and its velocity precisely. But on the quantum scale those things can't be pinned down; instead, they are represented by probabilities.* You can fence in a region where an electron exists, but as you tighten the boundaries of that fence, the electron behaves more erratically. The electron moves faster and more unpredictably. This is because you made its location more precise. The trade-off is that you can't know its speed as well. Nature does not like to be cornered.

Information about an electron is like a seesaw: On one side is its location, on the other its velocity. As you know more about one side, that end of the seesaw lowers, and the other end rises. The information about the electron is no longer balanced. Scientists would say the velocity and location (probabilities) are inversely proportional.

And so it is with the zone blitz. On a passing play, a defense does two things—pressures the quarterback and covers receivers. Coverage and pressure also tend to be inversely proportional. The defense has a finite amount of manpower, and the players who are back in coverage aren't pressuring the quarterback and vice versa. The zone blitz puts an almost quantum-like perspective on a passing play, shifting the uncertainty to the quarterback. When the passer focuses on the pressure, the coverage changes. When he focuses on the chang-

* The Uncertainty Principle actually states that the location, x, and the momentum, p, cannot be known precisely. Momentum is equal to mass times velocity, so we loosely use velocity in the text and assume that mass is fixed.

ing coverage, he's now a sitting duck for the pressure—that is, a blitzing defender looking to change his orientation from vertical to horizontal.

The zone blitz is designed to keep the offense from knowing in advance which players are rushing and which will drop back into coverage, known in the film room as "personnel exchange." Normally, a defensive lineman is too big and slow to cover a running back or a wide receiver in conventional man-to-man coverage, but even a 300-pounder can work as part of a zone scheme, which doesn't place nearly as much emphasis on foot speed.

Just as the no-huddle offense required versatile players who possessed a variety of skills—running, blocking, and catching passes—so it is with LeBeau's zone blitz. Versatile players—like safety David Fulcher and nose tackle Tim Krumrie in Cincinnati and safety Troy Polamalu and linebacker James Harrison in Pittsburgh—allowed LeBeau the flexibility to employ this demanding defense. Positions became less important than basic skills like tackling and coverage.

The conceptual implications of the zone blitz are profound. A good offense thrives on order and predictability. In much the same way that the West Coast offense, and later the no-huddle, kept the defense off balance, the zone blitz did the same thing to the offense. That uncertainty undermines the operational efficiency of the offense. A normal blitz takes one player out of pass coverage and allows him to rush. A zone blitz, in theory and in practice, allows any player to rush or drop into coverage. Which means that from play to play the quarterback doesn't know where the pressure will be coming from.

Modern variants on the zone blitz—like the amoeba, the psycho, and the colorfully named moving cow—try to confuse the offense further by taking defenders out of their conventional positions, sometimes leaving all eleven players upright before the snap.

And while no team plays the zone blitz on every play, the mere threat of it affects the way that a quarterback runs his offense throughout the game.

Seeing Cincinnati's zone blitz for the first time during the 1988 season must have been a bit like the British Army facing the minute-men during the Revolutionary War. Troops had been used to lining up and marching into battle in orderly formations to face a well-known enemy at a predetermined place and time. Now they had to fight insurgents hiding in trees and behind bushes, attacking when they least expected it.

"Penalties will kill you." That's what one knowing fan will say to another when his team gets called for pass interference. They'll say it because they've heard the coaches say it and the announcers repeat it.

And yet, it's not really true. If this adage were true, you'd expect that the best teams would commit very few penalties and the worst teams would collect flags by the dozen. That's not how it works.

Bad teams are penalized frequently on offense, while the best offenses don't get called for very many penalties. This makes sense, because of the nature of offensive penalties. The most common offensive penalties are procedure penalties. When an offensive player lines up a few inches into the neutral zone, moves a fraction of a second too soon, or commits some other small procedural infraction, it's probably the result of a lapse in concentration. Those same lapses in concentration tend to undermine the efficiency of an offense even on the plays when there is no penalty. The other major infraction on offense is the holding penalty, and desperate offensive linemen often resort to holding because they're physically overmatched—which is frequently the case when a bad team is struggling against a good one.

But on defense it's a different story. Believe it or not, the best teams often rank near the top in the number of defensive penalties. For example, in the 2011 regular season, the Baltimore Ravens allowed the third fewest points in the NFL, but they led the league in penalty yards with 1,032.

Why is that? Two reasons. The first is the structure of the game's penalty system. On offense, a holding penalty costs a team 10 yards, doubling the distance for a 1st down. More often than not, a mistake like that will kill the drive. But on defense it's a different story. Defensive penalties tend to be errors of aggression. And while a pass-interference call may seem damaging, it's not. A penalty generally leaves the defense in the same position they would have been in if the receiver had caught the ball—or even a better one.

Penalties exist at the boundary between a legal play and an illegal one. Or to look at it another way, they're on the surface of what's legal. In physics, surfaces are interesting because on the molecular level you have atoms that don't have all their needs met. They've run out of atoms for them to bond with, and as a result, surfaces behave weirdly. In some cases, surfaces have lots of energy associated with them because of those broken chemical bonds, which is why they're more reactive than other areas. That's why water forms into droplets, to reduce the surface area to an absolute minimum.

The surface tension that is created at these boundaries allows us to create bubbles that seem permanent and elastic, like rubber balloons, when they are actually temporary, a cousin to fog. Surface tension acts like a skin at the top of water, which supports walking bugs and floating paper clips, not to mention Monet's water lilies. Just as surfaces have special properties, so do those football plays that lie right on the boundary between a great play and a defensive penalty.

Let's use a real-world example from Super Bowl XLVII. It's late in the fourth quarter, and the Baltimore Ravens are leading 34–29. But the San Francisco 49ers have moved the ball inside the Baltimore 10-yard line. Their first three tries have been unsuccessful, but the score dictates going for it on 4th down. Niners quarterback Colin Kaepernick throws to Michael Crabtree in the right corner of the end zone.

Ravens cornerback Jimmy Smith has a choice: He can play a less aggressive defense and hope that Crabtree doesn't make the catch. Or he can go hard at the wide receiver and hope to keep him from making the reception.

If he plays less aggressively, hoping to avoid the penalty—and, possibly, the wrath of his coach and every fan in San Francisco—Crabtree makes the catch, and the Niners probably win the game.

But if he goes hard after Crabtree, there are three possibilities: Crabtree makes a spectacular catch despite Smith's best efforts. Smith prevents Crabtree from making the catch, Smith isn't called for the penalty, and the Niners turn the ball over on downs. Or Crabtree doesn't make the catch and Smith is called for pass interference. Even under this scenario, the ball is spotted at half the distance to the goal line, and while the Niners get an automatic 1st down, they still have to put the ball in the end zone. In two of the three scenarios, Smith prevents San Francisco from scoring.[*]

As you might remember, Smith went hard after Crabtree, who didn't make the catch, and the refs didn't make the call, preferring—as they say in the broadcast booth—to "let the players play." And thus a Super Bowl was decided.

The take-home lesson: The benefits for playing aggressive defense are large compared to the potential penalties.

In the 2012 NFC Championship Game, San Francisco 49ers punt returner Kyle Williams fielded a punt by the New York Giants. The ball was squibbing along on the ground, and Williams ran over to corral it. The ball had been rolling forward and suddenly, as Williams approached it, the ball rotated just a little. Then it took a random bounce and glanced off Williams's knee. The ball resumed its trajectory toward the Giants' end zone, and New York's Devin

---

[*] In college football the penalty for pass interference is different—a standard 15 yards and an automatic 1st down regardless of the length of the pass, meaning that the penalty for a 60-yard bomb is the same as for a 5-yard dink. Which encourages defenders to be especially aggressive on long passes, the exact opposite of the NFL.

Thomas' swooped in and grabbed the ball for a crucial fumble recovery. Williams would later fumble another punt in overtime, this one in a more conventional manner, for another crucial turnover that would result in the game-winning field goal. The Giants won the game and earned a trip to the Super Bowl.

Williams? He would get death threats.

At some level, that first fumble seemed like bad luck, a notable example of the kind of strange bounces that a football in the shape of a prolate spheroid will sometimes take.

Look a little closer, and it seems that this play was anything but random. A few weeks earlier in a regular season game against the Seattle Seahawks, Kyle Williams left the game after getting flattened on a return play. The team didn't disclose the nature of his injury, but after the game Williams informed members of the media that he couldn't talk to them, which is standard procedure when a player has a concussion or other head injury. Almost a month later, in that championship game against the Giants, Williams still didn't look at all like himself.

On that first fumbled punt against the Giants, he stared at the ball in a way that was downright strange for a professional football player. Williams positioned himself in a no-man's-land that most high school players know enough to avoid—not close enough to the ball to catch it, but not far enough away as to completely eliminate the possibility of getting tagged by a random bounce.

The Giants, for their part, seemed to be well aware that Williams was performing at less than 100 percent. "The thing is, we knew he had four concussions, so that was our biggest thing, was to take him outta the game," said Jacquian Williams, who forced the second fumble.

"We were just like, 'We gotta put a hit on that guy,'" said Thomas, who recovered both of Williams's fumbles. "[Giants reserve safety Tyler] Sash did a great job hitting him early, and he looked kind of dazed when he got up. I feel like that made a difference, and he coughed it up." These comments were reported after the game, and the postgame stories were archived online, but it gained little media traction at the time.

Some equally off-the-cuff comments made about Williams the week before would shake the football world to its core. In preparing his Saints for their NFC Divisional Playoff game against San Francisco, New Orleans defensive coordinator Greg Williams made some statements that, when they were made public weeks later, would push the NFL's Bountygate scandal—in which Saints players were allegedly paid cash to injure opposing players—toward its tipping point.

"We need to find out in the first two series of the game, the little wide receiver, number ten, about his concussion," said Greg Williams to his Saints defenders. "We need to f**king put a lick on him, move him to decide. He needs to decide." The little wide receiver? Number 10? That's Kyle Williams.

Greg Williams also talked about aiming for the head of running back Frank Gore, targeting the anterior cruciate ligament in wide receiver Michael Crabtree's knee, and performing similar mayhem on starting quarterback Alex Smith. As he peppered his pregame speech with aphorisms like "It's a production business" and "Respect comes from fear," Greg Williams sounded like a panicky sales manager giving his underperforming salesmen a pep talk as the quarterly deadlines approach. The tone suggested that Williams had been making these kinds of statements for years—this week's speech differing from last week's only in the way it highlighted the specific vulnerabilities of the upcoming opponents. As if the names were changed, but the message was familiar.

In the NFC Divisional Playoff game, the New Orleans Saints were faced with two options. They could have attempted to cover the almost countless options in the 49ers offense. Or they could hope that by putting a few starters on the sideline, having a few others sore or struggling with head injuries, and the rest thinking a little more about their own personal safety than the play at hand, they could level the playing field. They chose the latter option.

It may not seem right—and against the 49ers it didn't work—but the Saints won the 2009 Super Bowl with this strategy, and they were hardly the only defense in the NFL who employed it.

But there was one crucial difference with this particular speech: Greg Williams said this in front of a filmmaker named Sean Pamphilon, who was working on a documentary called *The United States of Football*. And when Bountygate became public, so did the recording of Williams's pep talk.

The release of the audio of Williams sparked a debate about overly aggressive defense and a possible link to the game's epidemic of head injuries.

To understand the scope of Bountygate, do the math. The Saints reportedly paid cash bonuses of up to $10,000 for knocking an opposing quarterback like Brett Favre out of the game. On the other hand, during the same period, the NFL fined players as much as $75,000 for illegal hits and doled out a suspension to Steelers linebacker James Harrison that cost him $215,000 in salary. After a helmet-to-helmet hit against Pittsburgh receiver Emmanuel Sanders, Baltimore Ravens safety Ed Reed was given a one-game suspension that would have cost him $423,529.41 in lost salary.[*] The NFL levied these fines after reviewing the videotape in the league offices in New York, often taking into account the injuries on the play.

Why would a player risk a $200,000 fine for a token $10,000 bonus? In an unspoken way, every defensive player in the NFL understood that his weekly paycheck was not entirely different from a bounty payment. This problem extends well beyond the bad judgment of one rogue coach and stems from the larger issue of overwhelmed defenses responding to complex new offenses in any way they can.

---

[*] The Pro Bowler appealed the suspension, and the one-game ban was lifted in favor of a $50,000 fine.

———

In 2008, *ESPN The Magazine* ran a short piece in which they asked Pittsburgh Steelers defensive coordinator Dick LeBeau and a few of his defensive players to riff on the A-11 offense. This complex scheme, used only in high school, employs two quarterbacks and makes every player on the field a potentially eligible receiver. How would they respond to this kind of complex attack in the NFL? Their candid responses are the product of an era before head injuries became a hot topic that demanded politically correct speech.

TROY POLAMALU: A high school uses two quarterbacks? Imagine for a second having Tom Brady and Peyton Manning on the same team, running this offense.

DICK LEBEAU: I'd pray for divine intervention.

POLAMALU: You can always hit offenses out of their schemes. So we'd play man coverage on the outside and put two beasts on either side of the line and blitz 'em from the edge on every down—you hit 'em, get the quarterback hurt. They start running out of quarterbacks, they won't play it anymore.

CASEY HAMPTON: Pressure busts pipes.

JAMES HARRISON: Yeah, this A-11 sounds dangerous to me. Dangerous for offenses. Guys spread out like that? All you have to do is shoot that big gap? I would love to see that in the NFL.

As defenses react to today's high-powered offenses and explore—and perhaps blur—the boundaries between tough defense and illegal hits, the beginning of what these Steelers predicted may be coming to pass.

PART THREE

THE PLAYERS

"I've been big ever since I was little."

—WILLIAM PERRY

CHAPTER 11

HOW TO TURN A BIG MAC INTO
AN OUTSIDE LINEBACKER

Ed Keenan deserves to be honored by the NFL. Nothing elaborate—after all, he played for only a year for the Hartford Blues almost ninety years ago. But something. A small plaque will do. In some team's weight room. Or maybe near the training table.

Keenan has the distinction of being pro football's first 300-pound player. Playing right guard for Hartford in 1926, he was listed at 6'4" and 320 pounds. For his day, Keenan was very much an outlier. It would be sixteen years before another 300-pounder, Green Bay's Millburn "Tiny" Croft, came along. Croft was listed at 300 pounds his first year but at 298, 285, and 280 in subsequent seasons—so he, too, was ahead of his time in a sport where fibbing about your weight has become as much a part of pro football culture as wearing eye-black.

Keenan was a trendsetter, as important in his way as, say, Jim Brown or Joe Montana, even if it took a while for the rest of the game to catch up to him. As late as the 1970s, according to a study by the Associated Press, there was only one 300-pound player in the NFL, San Diego's Gene Ferguson, and in 1980 there were only three. But in the mid-1980s, that changed: By 1990 there were ninety-four.

If there was an exemplar of this trend toward behemoths on the gridiron, it was William "The Refrigerator" Perry. A defensive lineman out of Clemson, Perry was a first-round draft pick of the Chicago Bears in 1985. Larger than life, literally and figuratively, the 335-pound Perry became an iconic figure on the Bears' Super Bowl championship team. ESPN's Tom Friend called him "America's mascot—a pear-shaped, gap-toothed football player who could sing, dance, sack quarterbacks, score touchdowns and muss Mike Ditka's hair." Despite his size, Perry could throw down a 360-degree dunk on the basketball court and was the sixth-fastest player on Clemson's football team. This combination of speed and mind-blowing size caught coach Mike Ditka's attention. And while Perry's play on the defensive line was spotty—which was a source of conflict between Ditka, who drafted Perry, and defensive coordinator Buddy Ryan—what made Perry into something of a folk hero was a series of brief appearances at running back. On six offensive plays, the Fridge scored three touchdowns and added another one in Chicago's Super Bowl XX win. And thus America fell in love with supersized football players, as NFL teams began to understand the way they could change the game.

Why are football players—and, for that matter, human beings—the size they are? Why, for example, isn't a defensive lineman eight feet tall? Why isn't a running back three feet tall?

Two words: *fire* and *falls*.

According to Ohio State economist Richard Steckel, who studies human height data, the answer can be found about three hundred thousand years ago in the middle Paleolithic period. Until that time, humans occupied a place somewhere in the middle of the food chain. We were skilled hunters, but we were also prey for large, fast animals, like saber-toothed tigers. The game changer was man's use of fire. Fire was the ultimate defense against virtually any would-be predator. It moved us to the top of the food chain. According to primatologist Richard Wrangham, author of *Catching Fire: How Cooking Made Us Human,* using fire to cook food also contributed to our larger brains and accelerated our evolutionary development.

The ability to make a fire conferred an important adaptive advantage on humans of a specific size. Imagine asking a three-foot-tall human—someone roughly the size of a six-year-old—to fell a tree, split it into manageable pieces, carry the logs over, and then pile the wood high enough to make a fire that will last all night. It's not hard to see that a human being closer to six feet tall is better suited to all of those tasks. That, according to evolutionary biologists, is the best guess as to why humans aren't significantly smaller than we are.

But then why aren't we *eight* feet tall? We can lay this one at the foot of geometry. In 1638, Galileo published a book called *Dialogues Concerning Two New Sciences* in which he argued that the size of a human being is far from arbitrary. He posited something called the square-cube law, which suggested that volume—and hence mass—increases rapidly with small increases in height.

For a human, that means that if you increase a person's height, his or her mass will increase significantly. Evolutionary biologist J. B. S. Haldane, author of the 1926 essay "On Being the Right Size," argued that "comparative anatomy is largely the story of the struggle to increase surface in proportion to volume."

Using Perry as an example, if we double his height from 6'2" to 12'4", his weight would increase from 335 pounds to 2,680 pounds. By comparison, a Toyota Yaris weighs about 2,300 pounds, so a supersize Perry would weigh as much as a small car with two passengers in it.

And that can contribute to serious falls. Remember how you tumbled as a toddler and then giggled as you got up? The same kind of fall might send an adult to the emergency room. Beyond clumsiness, there are other problems with gigantism. Even a rather modest increase in height would require a major overhaul of our physiology.

The huge increase in mass would tax the ability of our bones to support that weight. It would challenge our lungs to absorb enough oxygen to feed all these tissues. And it would require a high-pressure circulatory system to transport that oxygen to the cells. In the few people who are more than a foot above average, one or more of these problems tends to crop up, and they often battle a variety of chronic and acute disorders.

That's the reason why you simply don't find old giants. Or eight-foot-tall quarterbacks.

While William Perry's charisma brought plus-size players into the NFL limelight, there are practical reasons why so many contemporary football players are 300 pounds or close to it. The game has evolved.

As we mentioned previously, in 1978 the NFL enacted a series of sweeping rule changes designed to open up the offense. One deceptively significant tweak in the rules addressed the way offensive linemen could block: They could now use their hands. Before 1978, players were forced to block with their bodies, leaving their arms at their sides like chicken wings. They resorted to chest-bumping opposing defensive lineman or driving with their shoulders. "They'd just chug along with short, choppy steps and try to block people," explains former Cincinnati Bengals coach Sam Wyche.

When forced to block this way, foot speed was every bit as important as sheer size. A supersized offensive lineman who wasn't nimble enough to keep his bulk between the defensive end and the quarterback wasn't much good.

"They changed the rules so you could extend your hands straight out in the framework of your body," Wyche recalls. "As soon as this happened, you saw more screens, you saw more sweeps, because now you could pull that big 290-, 300-pound man out of the interior line, and he could still block that nifty 185-pound defensive back, because he could shove him. He didn't have to get that close to him to affect him. Before that, those little defensive backs would dance around those linemen. They were useless out in the open field. So the rules do make a difference."

The new rules worked in yet another way to pave the way for large blockers: They actively encouraged the passing game. Pass blocking is fundamentally different from the run blocking that Vince Lombardi so lovingly diagrammed. On a running play, the offensive linemen move forward, creating holes for running backs, and once they've done that, they continue down the field to intercept other would-be tacklers. Run blocking is a more active discipline and requires a more agile athlete. And, generally, a smaller one.

On a pass play, the blocker's job is to keep the rush at bay long enough for the quarterback to find an open receiver. Offensive linemen in a pass-blocking scheme absorb the energy from the rush in a largely stationary position. "They're working in a phone booth," explains Wyche. If a lineman who's run blocking is an irresistible force, a lineman in pass-blocking mode is the immovable object. Pass blocking takes just as much skill as run blocking, but a different kind. And it requires a different body type—a bigger one.

With that development, fast hands became just as important as fast feet, and—as *Vogue* editor Anna Wintour might say—big was the new black.

So in the 1980s, at the same time that the United States and the Soviet Union were beginning to wind down the nuclear arms race, NFL teams started in earnest on a race of their own. With an increasing emphasis on pass blocking, offensive linemen ballooned in size. Here's a list of some Hall of Fame offensive linemen from different eras, with their listed playing heights and weights.

Frank Gatski	1946–1956	6'3"	233
Forrest Gregg	1956–1971	6'4"	249
Dan Dierdorf	1971–1983	6'3"	275
Bruce Matthews	1983–2001	6'6"	289
Willie Roaf	1993–2005	6'5"	300
Jonathan Ogden	1996–2007	6'9"	345

Ogden, who played for the Baltimore Ravens at the turn of the millennium, is 48 percent larger than Gatski, who was a fixture of the Browns' championship teams of the 1950s. And as the offensive linemen grew, so did the defenders who were charged with fighting off their blocks on running plays or getting around them on pass plays.

The battle of the bulk continues to this day. As we mentioned earlier, in 1980 there were three 300-pounders in the NFL. By 2012, that number had grown to 461.

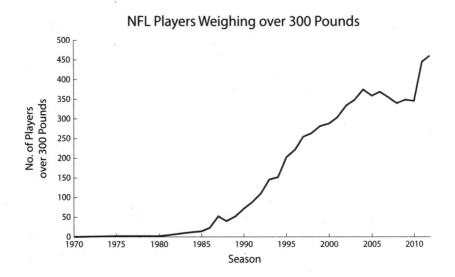

NFL Players Weighing over 300 Pounds

"You are what you eat."

We all buy that on some level.

But look at a Big Mac. And look at Ray Lewis. And look back at the burger. It's easy enough to toss around these nutritional truisms, but how—on a granular level—do two all-beef patties special sauce lettuce cheese pickles onions on a sesame seed bun turn into a future NFL Hall of Famer?

"One of the great human desires is alchemy," explains David

Katz, founding director of the Yale-Griffin Prevention Research Center. "While we all secretly lust for the ability to turn one material into another, that's just what our bodies do every day."

And so does every NFL linebacker. Katz likens growth in our bodies to an ongoing construction project, with our food being the raw materials.

"Most of our homes are constructed largely of wood," he explains. "And we extract wood from trees. In the same way, the body extracts nutrients from food." The raw materials in food must be reduced to chemical building blocks, and that begins the moment we take our first bite. "The chemistry begins in our mouth," Katz explains. "There are enzymes in saliva like amylase and [salivary] lipase that are designed to target starch and fat at particular junctions and break down complexes of starches and fats into smaller units. That continues down into the stomach and the intestines, and by the time that chemical digestion is done, essentially what you've got is fat molecules and protein molecules and carbohydrate molecules."

But how exactly does the burger turn into a human being? The bun is made largely of starch. So the linebacker converts it into simple sugar. That simple sugar is made into glycogen, which is stored in the liver along with some smaller amounts of unconverted starch. These glycogen stores are the body's short-term energy source. The body uses these sugars to power all sorts of things: the muscular contractions that a player uses in executing a tackle, the neurological processes at work when he's listening to his coach talk about the new zone blitz package, and even the very process of converting that Big Mac bun into glycogen.

If Ray Lewis is doing two-a-day workouts at training camp, all of this glycogen is used to power his body, with little or nothing left over. However, if it's a day in the off-season when the only football he's playing is some version of Madden and he's primarily exercising his right thumb, then he'll end up with a net caloric surplus. That glycogen will move from short-term storage in the liver to long-term storage as fat in various cells.

"It's subject to the laws of thermodynamics," says Katz. "If you don't burn it, the body will store it as an excess of calories."

The fat? Some of it will be used in the walls of every cell. But beyond that it'll be stored as, well, fat. Katz explains that, in a mature individual, the number of fat cells don't change except under the most dire of circumstances. The total quantity of fat cells is set during childhood and adolescence, and in adulthood the fat cells merely change in size, growing when there are extra calories, decreasing when there's a deficit. Only in truly obese individuals do you start to see the creation of new fat cells, a condition called hypertrophic obesity.

The protein in the burger will break down into amino acids, which are used as the building blocks for new muscle cells. And unlike fat cells, the number of muscle cells will increase as the player hits the weight room. An NFL player builds muscle on an ongoing basis. But any excess protein, Katz notes, also becomes fat.

Why is the body so remarkably good at storing fat? Again, the answer is evolution. Cheap and easy access to calories is a very recent development in the human condition. The hunting and gathering that early man did was a boom-or-bust business. One day there'd be a temporary feast in the form of ripe fruits and vegetables or a freshly killed ox. But there were, of course, no Ziploc bags or Sub-Zero refrigerators in which to store the leftovers.

When the harvest was over and the hunters hit a dry spell, it was famine time. Attempts to store food were generally unsuccessful, and even when they did work, someone would have to defend the food stores against those who'd steal them, human or otherwise.

Storing excess calories as fat was an elegant solution to these problems.

"Fat is the best defense against a rainy day, and throughout human history there were lots of rainy days," Katz explains.

Katz is quick to point out that a Big Mac is far from an optimal meal. "It's poor-quality fuel," he explains. "One of the problems we get into is not respecting the limits of this practical alchemy. If you put in bad fuel, it wears out the components."

This, of course, addresses the complex phenomenon of building a body over the course of just one meal. But over a longer span, it gets to be even more of a head-scratcher. Take Ray Lewis on the day he played in his first Super Bowl in 2000 and Ray Lewis on the day he played in his last Super Bowl in 2013, and you could argue that he's not the same man at all. And you'd be right: Almost all of his cells have been replaced with new ones.

Katz draws an analogy between our bodies and a river: The individual water molecules are constantly changing, but the size and direction of the river change slowly, if at all.

"Our bodies are much more like a river than a mountain," he explains. "We replace all the cells on the surface layers of our skin and our intestinal tract in a matter of days to weeks. Red blood cells live about a month. Over a lifetime we replace all the cells in our bodies multiple times over. We're constantly flowing."

Or maybe it's more like dinnertime at the Corleone compound. "The way that a given body puts cells together is unique to that body, but it's the same basic set of instructions, the same recipe," Katz explains. "If you're making spaghetti and meatballs, you can double the recipe, or if you've got a group that's really hungry and carnivorous, you can keep the amount of spaghetti the same and double the meatballs. But it's the same basic dish."

And then there's the question no one wants to ask: Are a significant number of those 300-pounders in the NFL going beyond the Big Mac and taking steroids or other performance-enhancing drugs to bulk up?

Talk to Dr. Charles Yesalis, author of *The Steroids Game,* and his answer is a resounding yes. "God doesn't readily change the recipe," he explains. "It takes a tad more time than a decade or two. Clearly, performance-enhancement drugs have altered the equation."

Professor Jay Hoffman, a former NFL linebacker and himself a former steroid user, contends that steroid use is on the wane. "I would have agreed with him twenty years ago. I don't think steroids are used as much today."

What exactly are steroids? They're essentially synthetic testosterone, which is the sex hormone that makes men men. They build muscle, and they are used to counter bone loss for those suffering from osteoporosis. They're also used as anti-inflammatory drugs.

When Yesalis interviewed elite strength athletes from the 1950s and 1960s, he found that most of them peaked at a lean 230 pounds using only resistance training. When the athletes started using steroids and growth hormones, those same athletes got to "270, 280, 290, and above," he explains. "Their training techniques were remarkably similar to the ones we are using today. If you want to increase the strength of your triceps, you add resistance against straightening your arm out. It isn't rocket science," he argues. "These drugs will take you to places that you'll never get to naturally."

Hoffman, who used steroids for several years during his brief pro football career with the Jets and the Eagles, exhibited similar gains in size and strength. "I was 245 lean, and then ten weeks later I'm 275," he explains. How would he have fared without steroids? "In a ten-week time frame I could have put on 10 pounds of lean tissue with the supplements that are out today. Maybe without steroids," he estimates, "I would have gotten to 250 to 260. But not as lean or as strong."

Hoffman further explains he made those steroid-enhanced gains with almost no additional weight lifting. "I never really lifted many weights," he says. "I never lifted much in high school. I just played lots of sports. I played the seasons. I played basketball in the winter, football in the fall, and baseball in the spring. I played lots of school-yard basketball."

He suggests that today's pro football players are making the same kinds of gains that he made using steroids through more comprehensive strength training. "We are training better, there is more knowledge about supplements and resistance training," he says. "There are more supplements that help athletes achieve their goals, and more coaches to help them achieve their goals."

Hoffman and Yesalis agree that most of the 300-pounders in the NFL are content to add sheer mass as well as muscle. Offensive linemen in the league sport a body mass index that puts them on the wrong side of obesity. So they're not only big, they're fat.

For Yesalis, the incentives to use steroids seem too great for many players *not* to use them. The motive and the opportunity are all too obvious. "There are huge amounts of money, fame, and sexual rewards," he explains. "These drugs, as well as the scientific backup to evade drug testing, are readily available to elite athletes."

While he's convinced that steroid use is prevalent among professional football players, Yesalis isn't overly worried about its effects on the players. He notes that steroids and other performance enhancers have been used as therapeutic drugs for decades, and that—used with reasonable care—they're relatively safe.

"These are not major killer drugs. No epidemiologist would put them in the same league as methamphetamine or tobacco or cocaine," he argues, while positing a thought experiment. "If I put in front of you a big bottle of Valium, a bottle of acetaminophen, and a bottle of steroids, and held a gun to your head and told you that you have to ingest one of these bottles or I am going to blow your brains out, you'd better damn well take the steroids, which will give you an upset stomach. The other two will kill you."

While he's more concerned about steroids than Yesalis is, Hoffman sees another problem facing NFL players: pain medication. "Masking injury is probably the most dangerous thing we can see in football players," says Hoffman. Before and after games athletes routinely take medications like Toradol and other potent anti-inflammatory drugs. He notes that these drugs can be dangerous in and of themselves, and furthermore, they put athletes at risk for long-term injury by covering up important pain signals. "The public looks upon a player who is willing to take medication to mask the pain as a hero," Hoffman says, "and someone who wants to maximize his performance a cheater."

Just how big is the problem of performance-enhancing drugs in the NFL? Since 2010, there have been fifty suspensions for all banned substances. Given that there are about 1,700 players in the league, that relatively small number can represent a glass that's half empty or one that's half full. A rather large number of players may be using performance-enhancing drugs and most of them are getting away with it. Or perhaps testing has proven to be a major deterrent and few players are using drugs because of the risk of getting caught and suspended. Just like a police department's surprisingly low arrest numbers, these stats can be seen either as evidence of a system that's working very effectively or one that's seriously flawed.

In any case, one of the world's foremost experts on the subject of performance-enhancing drugs remains largely unconcerned. "I purchased NFL Ticket this year, so I can watch my Steelers," says Yesalis. "It is just that I take the whole damned thing with a block of salt, not a grain."

CHAPTER 12

SIR ISAAC NEWTON'S
FANTASY FOOTBALL DRAFT

It's the most clichéd of all interview questions: *If you could have dinner with any person, living or dead, who would it be?*

"Einstein."

"Elvis."

"Kate Upton."

Can we just skip to the question about what kind of tree I'd be?

Let's instead spin this chestnut of a question in a more interesting direction: Who would you want running your Draft Day war room? Upping the ante, let's imagine it is the Ultimate NFL Draft: It would work a little like a fantasy draft in a keeper league, except you'd be picking among not only this year's crop of rookies but also *all* rookies going back to the year 2000.

Who would you want to advise you? Some would go for legendary Pittsburgh Steelers coach Chuck Noll. Others might pick Bill Walsh. Still others would adopt a WWJD strategy: Who Would Jesus Draft?

For our part, we would like to suggest Isaac Newton.

He's a classic out-of-the-box thinker. He'll crunch the numbers. And the data from the NFL Combines fits neatly into his equations.

So, among recent vintage NFL players, who is Isaac Newton's first-round pick? First, a little physics.

$$p = mv$$

As equations go, it might not be quite as sexy as $E = mc^2$, but it is not only at the root of modern physics; it's also at the root of modern football: m is mass, v is velocity, and p is momentum.

But what *is* momentum? Newton himself called momentum the "quantity of motion." A high school physics teacher might call it "mass on the move." Whatever it is, it's indestructible. And while an eight-year-old might get jazzed about the fact that it can't be destroyed, a physicist latches on to the even more profound concept: Momentum is *conserved*. Which means that every bit of momentum must be accounted for. If a moving object hits an inert one, the momentum of both objects—which are now moving—must equal the initial momentum of the first object. The energy can change form— turning into heat generated by friction, for example—but it *must* be accounted for. Unlike that sock you lost in the dryer, momentum has nowhere to hide. Momentum in equals momentum out.[*]

What sports fans like about momentum is this: It can be measured. Total momentum depends on two things: the amount of stuff and how fast it's moving. In football terms, size and speed.

Speed and size are the yin and yang of professional football. Both are highly desirable, but they tend to be found in inverse proportion to each other. The player with speed is generally lacking in size, and vice versa.

In professional football, momentum is a way to combine both of

[*] Since momentum is velocity multiplied by the mass of an object, the units are mass-speed—or lbm-ft/sec. The unit *lbm* means pounds-mass; it is used to differentiate from pounds-force, which is denoted as *lbf.*

these crucial attributes into a single number, a number that translates into a real-world skill that every player uses—the ability to hit another guy hard while running at top speed.

The bigger the number, the harder the hit.

So what would Special Assistant Isaac Newton whisper in the general manager's ear on Draft Day? Like all the other scouts and player personnel guys huddled around their screens as the teams try to clarify their draft boards, Newton would have access to a dizzying array of data, ranging from game films of a player's best and worst plays to interviews with his high school coaches and his neighborhood friends.

But Newton would want only two numbers.

A player's weight.

And his time in the 40-yard dash.

A few hundredths of a second made running back Chris Johnson a millionaire in the 2008 NFL draft. Following his senior season at East Carolina University, Johnson was on the bubble. If the picks fell right, he'd be happy being drafted in the second round. If they didn't, the NCAA record holder in all-purpose yardage might have dropped to the third round or lower, watching helplessly as player after player got picked.

And then came the 40. Johnson arrived in Indianapolis for the NFL Combine, where players assemble for speed, strength, and agility drills. His dreadlocks bouncing in the breeze, Johnson ran the 40-yard dash in 4.24 seconds. His time was—and remains—the fastest electronically timed 40-yard dash in the history of the NFL Combine. He came to Indy hoping to simply outrun his rival, Arkansas back Darren McFadden. Instead, he made history.

How long is a hundredth of a second? A blink of an eye is about 35 hundredths of a second. Pick up a stopwatch and try to start it and stop it as fast as you can. Even with practice, your best times will be between one and two tenths of a second, or about 15 hundredths of a second.

Hundredths of a second: Human beings simply don't work that fast. Which is why, in the early history of track and field, major competitions were hand-timed to the tenth of a second. There were experimental electronic timings as far back as 1928, but they were unofficial. Not surprisingly, there is a substantial difference between the human eye and the electronic eye. The 1964 Olympics marked the first time that fully automatic time—FAT for short—was used; even then the officially posted times were rounded to the nearest tenth of a second to make them consistent with earlier hand-held times. Bob Hayes won the gold medal in the men's 100-meter dash with a time of 10.06 seconds—which was posted as an official time of 10.0.* Officials hand-timing the race with stop watches recorded a time of 9.9 seconds. It wasn't until 1977 that the International Amateur Athletic Federation (IAAF) required electronic timing to the hundredth of a second for world records.

Chris Johnson's run—as all 40-yard dashes at the Combine since 1999 have been—was timed with a semiautomatic system. Faster times by Bo Jackson (4.12) and Deion Sanders (4.21) were fully hand-timed, which produced times that were less accurate—and significantly faster—than the current method in which the clock is hand-started but stopped electronically. Johnson's time tied that of Rondel Menendez of Eastern Kentucky, who also ran a 4.24 in 1999, the first year of the current electronic timing system. All in all, while the Combine's timing is more than adequate to separate the fast guys from the slow guys, the system is less accurate than what you'd find at your average high school track meet. Which is ironic, given that teams are making million-dollar decisions based on this data.

Still, with that one warp-speed sprint, Johnson caught the attention of NFL GMs and player personnel directors. In just over four seconds, Johnson vaulted himself from a marginal second-round pick to a first rounder on the rise. He was selected with the twenty-fourth pick by the Tennessee Titans. Had Johnson run, say, a tenth

---

* Hayes's actual time was 10.01, but the system had a built-in delay of 0.05 seconds, so track-and-field scholars use Hayes's 10.06 when comparing his times to more modern efforts.

of a second slower at 4.34, he probably would have remained a middle-round selection, and it's possible that the team that drafted him would not have given him the opportunity to shine usually afforded to first-round picks.

"If I had run a 4.30, I would have been upset," Johnson said before the draft.

Or, as his agent Joel Segal explained, "You're talking millions of dollars for a tenth of a second."

Instead, that blazing 40 helped make Johnson an NFL star. The Titans signed him to a five-year, $12 million rookie deal. And he quickly exceeded even Tennessee's expectations. In his first year he rushed for 1,228 yards, finished second in the Rookie of the Year voting, and was selected to the Pro Bowl. In his second year, he rushed for 2,006 yards, becoming just the sixth back ever to rush for 2,000 yards in a season. His 2,509 yards from scrimmage set an NFL record. Before the 2011 season, Johnson inked a four-year, $55.3 million contract extension—with $30 million guaranteed—which briefly made him the highest-paid running back in NFL history.

Johnson was the poster boy for pure speed in the NFL. He was also quite the exception. Contrast his career arc with that of Rondel Menendez of Eastern Kentucky, the wideout with whom Johnson shares the Combine record. Menendez was drafted in the seventh round and never caught an NFL pass. Wide receiver Darrius Heyward-Bey ran a 4.25 and cornerback Fabian Washington clocked a 4.29, and both were selected in the first round by the speed-obsessed Oakland Raiders, but also they haven't lived up to expectations.

The secret of Johnson's success? Work. Track work, to be exact. While it's not widely known, virtually every top NFL draft prospect begins training like a track star as soon as he hangs up his football cleats in January. Johnson spent eight weeks working with NFL speed guru Tom Shaw, who runs an independent Combine preparation camp near Orlando, Florida. At the camp, which cost Johnson $750 a week, athletes take part in grueling two-a-day workouts, building strength by pulling parachutes and stretching bungee cords and improving leg speed by running downhill sprints, during which Johnson clocked a slope-aided time of 3.6. Johnson was already blazingly fast when he arrived in Orlando, but Shaw made him even faster by increasing the length of his stride. "If you add two inches to each stride, that's forty inches at the end of forty yards, or about two tenths of a second," Shaw explained. At the beginning of Shaw's camp, Johnson needed 19 strides to run a 40. He was down to 18 when he set the record. That one step pared away made Chris Johnson an NFL superstar. And a very rich young man.

Speed merchants like Mr. Johnson aren't the only ones whose Draft Day fortunes rise and fall with the stopwatch. While Johnson's stock was on the upswing in the 2008 draft, 300-pound defensive lineman Sedrick Ellis of USC was headed in the opposite direction. After running a disappointing 5.26 at the Combine, he knew he needed to get fast in a hurry.

The time was slow enough to raise questions about whether Ellis, who moved plenty fast when chasing opposing backs, was too slow for the NFL. Despite a stellar résumé as part of USC's powerhouse pro-style squad, Ellis and his agent, Eugene Parker, began to hear whispers about his speed. "NFL general managers, are like loan officers," said Parker. "They're always looking for a flaw." If Ellis dropped out of the top 10 of the draft to the middle of the first round, it would cost him big money.

It turned out that Ellis had almost as much riding on his sprint as Johnson.

Ellis checked in for a crash course at a competing Combine prep program at Saddlebrook Resort near Tampa. According to Jason Riley, Saddlebrook's director of performance, Ellis was a disaster on the track. He popped up quickly after his start, then folded at the waist as he lumbered down the track. His arms were flailing, his torso was twisted, and his strides were slow and long. Riley even noticed that Ellis dropped his toes as he ran. And all of these flaws were exacerbated by his sheer mass. "He was spending a lot more time on the ground than he needed to," Riley explains.

Riley's program consisted of single-leg exercises to increase Ellis's explosiveness followed by stretching and massage to increase the range of motion in his hip and ankle joints. From a technique perspective, Riley deconstructed Ellis's running style, teaching him piston-like steps at the start and more relaxed strides once he got under way. During his time at Saddlebrook, Ellis also lost weight, going from 315 to an even 300. His time of 5.02 in his USC Pro Day silenced his critics. He nipped the Pro Day time of rival Glenn Dorsey from LSU and again solidified his status near the very top of the draft. Ellis was drafted by the New Orleans Saints, who traded up to grab him with the seventh pick, and he would become a starter on a Super Bowl team. An opportunity that presented itself because of two tenths of a second in a 40-yard dash.

Johnson and other NFL speed burners owe a special debt of gratitude to Bob Hayes, who helped ignite the league's obsession with speed.

One of the greatest sprinters of all time, Hayes broke the world record in winning the 100-meter dash in the 1964 Olympics—where he also earned the unofficial title as World's Fastest Man—and anchored the U.S. world-record-setting, gold medal 4 × 100 relay. But the Olympics would be his last meet as a track star.

The Dallas Cowboys drafted Hayes in the seventh round of the 1964 NFL draft. A late-round pick like this was usually considered almost an afterthought; Bill Parcells was selected one pick after Hayes in that draft by the Detroit Lions, and while he's a Hall of Famer coach, as a player he was cut without ever playing a single game.

Hayes, on the other hand, was anything but a wasted pick. He had been a high school star but didn't play at Florida A&M, so his football skills were rudimentary. During the 1967 NFL championship game—also called the Ice Bowl—Hayes evidently tipped off the

Green Bay Packers' defense on that frigid afternoon by putting his hands in his pants to warm them—but only on running plays, when he knew he wouldn't be called upon to catch the ball.

But there's no substitute for speed. Hayes would use his world-class quickness to lead the league in receiving touchdowns his first two seasons. He would catch only 371 passes in his career, but he averaged an astonishing 20 yards per catch. As it became clear that even the fastest cornerbacks couldn't cover Hayes one-on-one, defensive coordinators began using schemes like the bump-and-run to slow Hayes down, and deep variations of the zone defense to contain him.

"I doubt that there has ever been anyone who revolutionized the offensive game the way Bobby did," said his teammate Don Meredith, who was Hayes's first pro quarterback. "His amazing speed forced the defense to do a complete reevaluation of what it had to do to stop him."

As the moments tick away in this fantasy Ultimate NFL Draft, Team Newton is about to get its turn on the clock. Knowing a lot about physics but little about football, Sir Isaac has a plan. He decides that the player he'll select is the one who can generate the most momentum. He goes off to plug those numbers into an Excel spreadsheet on his Apple computer.

Which player will Newton select? Will it be a player with insane speed like Chris Johnson? Will it be a wide receiver like Calvin Johnson (no relation), who's a tick slower than Chris at 4.35 but bigger at 239 pounds? What about a defender? A linebacker like 242-pound Patrick Willis of the 49ers, who ran a 4.51 40-yard dash, or better yet, a defensive end like 266-pound Dwight Freeney of the Colts, who put up a 4.48?

It's the sort of question that football fans could spend all night arguing about—over a beer, probably, instead of a laptop—with some leaning toward straight speed or size and others looking for a balance.

Newton exhibits no such uncertainty. Using his momentum calculations, he identifies a winner quickly and easily. It isn't even particularly close.

Isaac Newton's first overall draft pick in the Ultimate NFL Draft? Aaron Gibson.

Who, you might be wondering, *is that?* Gibson was a 6'7" offensive tackle out of the University of Wisconsin who went on to play for five years in the NFL, with the Cowboys, the Lions, and the Bears. Gibson ran only a 5.4 40-yard dash, but he weighed in at *386 pounds* while doing so. Which means that Gibson could produce momentum of 8,662 lbm-ft/sec.

Next on Newton's list is 374-pound Derrick Fletcher, who ran a 5.2, followed by Dontari Poe, who ran a 4.89 at 346 pounds. Rounding out the top five, T. J. Barnes and Leonard Davis. Detect a pattern here? These are all very, very large players. The first sub-300-pounder on the list is Mario Williams, who weighs 295 pounds and is ranked 120th on Newton's Momentum List.

Gibson is not merely a fast big guy. He is flat out the biggest player on this list, probably the biggest player ever to play in the NFL, weighing 440 pounds at his peak. When he was a senior in college at Wisconsin, Gibson's *thighs* measured 33¼ inches, which is more than 5 inches bigger than the *waist* of one of his Badger teammates. His head is size 8⅞, which required a custom-made helmet, and his 18 EEEEE shoes are so big that his mother would borrow them to wear to her part-time gig entertaining at kids' parties as a clown.

Where does Chris Johnson, the fastest guy ever, rank on the Momentum List? Running a 4.24 40-yard dash at 197 pounds, his momentum value is 5,380 lbm-ft/sec, which is only 62 percent of Gibson's momentum. Among 3,841 NFL prospects since 2000, Johnson ranks 2,592nd.[*]

---

[*] Bill Barnwell at Football Outsiders used a player's weight and his 40-yard-dash time from the Combine to create a stat called Speed Score, a tool for evaluating running backs entering the NFL. The stat has proven to be a useful metric for predicting a player's performance as a pro; Johnson's speed score of 121.9 was better than the more highly touted Darren McFadden.

What do these momentum numbers actually mean?

Imagine a game of football played on a frictionless field. Put Aaron Gibson 40 yards away and let him run as fast as he can and throw a block at 200-pound safety Ed Reed. If Gibson could stop immediately and transfer all his momentum to the stationary Reed, the smaller player would slide away at 21 mph. Run the same experiment with Johnson doing the hitting, and Reed's speed would be only 14.8 mph.

These rankings may seem counterintuitive. A player who's both big and fast—in the mold of Lawrence Taylor, say—would seem like the obvious choice. And in an actual football game, Gibson is less perfect than on Newton's chalkboard. Momentum is directional, so while the amount of momentum he can produce is impressive, redirecting that momentum—that is, turning and stopping—becomes a problem.

Despite that, if we examine the numbers more closely, Gibson's supremacy still makes a certain kind of sense. Gibson's running speed is 15.29 mph. Chris Johnson's is 19.29 mph. Johnson is 26 percent faster than Gibson. Gibson weighs 96 percent more than the 197-pound Johnson.

To look at it another way, Gibson is more than twice as massive as the smallest cornerback, while Johnson is nowhere near twice as fast as even the most glacial big man. That's the reason why the massive Gibson ranks first, and Johnson, his blazing speed notwithstanding, ranks toward the bottom.

Turns out that NFL coaches and general managers are pretty good intuitive physicists. Over the last fifty years, there's been virtually no increase in top speed among the fastest players. Chris Johnson isn't meaningfully faster than Bob Hayes. Speed is important, but it has become almost a constant.

Size is not. The biggest players are almost 100 pounds heavier than they were in the mid-1960s. And as we learned in chapter 11, there are far more 300-pounders in the NFL than ever before. When we look at NFL rosters and the corresponding salary caps, these two facts snap into sharp relief. While Chris Johnson's attention-getting 40 likely made him a star, he was the exception that proves the rule. Given the tiny gap between the fastest players and those just a tick slower, NFL teams view speed as a commodity that's easily replaced. That's why a player like Rondel Menendez, who shared the Combine sprint record with Johnson, had such a brief and anonymous pro football career. It's not that speed isn't important; it's just that it's easy to find.

Size, on the other hand, is prized. NFL teams look at a player's mass the way NBA clubs look at height: More is better. And teams are willing to pay for it. As Michael Lewis notes in his book *The Blind Side,* the giant left tackles assigned to protect the quarterback trail only those very QBs on most teams' salary caps. And while Gibson had a less-than-stellar NFL career, he actually received many more chances—he played with four NFL teams—than a more modest-sized player with his football skill set and injury history would have received.

As he was crunching those momentum calculations, Newton was echoing one of pro football's fundamental truths: the bigger the better. Of the twenty-five highest-paid players in the NFL in 2012, seven were quarterbacks. Of the remaining eighteen, only six were speed guys: wide receivers, running backs, defensive backs. The remaining dozen were offensive or defensive linemen—big guys paid handsomely for their ability to bring momentum to the football field.

CHAPTER 13

CHOOSING YOUR NEXT QUARTERBACK? THEY HAVE AN APP FOR THAT

Nowhere is the gap between what NFL teams want and what they get more profound than when it comes to drafting a franchise quarterback. It is generally agreed that the quarterback is the most important player on the field, and that's reflected in everything from the size of their salaries to the rules the league has enacted to prevent them from getting injured.

And yet, despite the clear importance of finding the right quarterback, teams do a conspicuously poor job of drafting them. Since 1990, twelve quarterbacks have been taken with the first overall pick in the NFL draft. Of those picks, only two—Peyton Manning and his brother Eli—can be considered unqualified successes. (Recent draftees Andrew Luck, Cam Newton, and Matthew Stafford have shown promise early in their careers but are not yet locks for long-term stardom.) Other top-drafted quarterbacks—Tim Couch, JaMarcus Russell, David Carr—have been flat-out busts. And entering the 2012–13 season, those top-pick quarterbacks not named

Manning had posted a composite record of 400–509–3 for a .440 winning percentage.

By contrast, several All-Pro and Hall of Fame quarterbacks weren't even chosen in the first round, which means that virtually every team passed on them at least once. Drew Brees, who led the Saints to a Super Bowl win, was chosen in the second round, as was Canton-bound Brett Favre. New England's Tom Brady was chosen in the sixth round of the 2000 draft—and he boasts a career winning percentage of nearly .800.

Why do teams find it so hard to make good decisions when drafting a quarterback?

"The quarterback position is such a difficult position to evaluate because it's part physical talent and part cerebral talent," says Paraag Marathe, chief operating officer of the San Francisco 49ers. "The standard deviation in physical skill set is so small at the NFL level. By the time they get to the NFL, players have been weeded out physically ten times over—starting at Pee Wee and high school; in Division III, Division II, Division I; being a starter versus a redshirt player; getting drafted in the first round or the seventh round," Marathe continues. "But mentally, there's a bigger gap. That's the difference between a successful quarterback and an unsuccessful quarterback."

Charlie Ward is uniquely qualified to explain the mental demands of the position, having led teams in both pro basketball and college football. Playing quarterback for Florida State's 1993 national championship team, he won the Heisman Trophy in one of the biggest landslides ever. But despite his success, pro stouts were wary of Ward's modest size (6'2", 195 pounds) and his less-than-stellar throwing arm. Teams focused on his body rather than his brain. Draft Day came and went, and not one NFL team drafted the reigning Heisman Trophy winner. Instead of playing pro football, Ward opted for the NBA, where he played point guard for a strong New York Knicks team, which was a perennial playoff contender.

To put Ward's explanation of the cerebral side of quarterbacking into a scientific context, we enlisted Charles Greer, a Yale University brain scientist.

WARD: Clarity of mind is very important for a quarterback or a point guard.

GREER: I can appreciate the importance of clarity of mind, if by that he means being able to focus on the job at hand. Not being distracted by the crowds and the cheerleaders. A quarterback like Ward has to focus on two things: the people trying to break his legs and the guy running down the field he wants to throw the ball to. Other than that, he doesn't have anything to worry about. *(laughs)*

WARD: Decision-making on the field or on the basketball court is a little like driving. You have to make instantaneous decisions about whether you're going to change lanes. It's similar out on the field. You have a pre-snap read. You have a post-snap read. And then you have the actual play. And while all that's going on, there are people moving, and you have to make a decision about where you want to throw the ball.

GREER: To execute a complex task like this, Ward doesn't have to do things simultaneously. He can do them sequentially. He can focus first on how effective his line is at protecting him. Then, after a second or two, he can shift his attention to where his receiver is, and at that point he can throw the ball.

If you're a musician, you don't think about Beethoven's Ninth in its entirety when you're playing it. You just play one note at a time. Collectively, it's extremely complex. But if you look at one note at a time, it's no big deal.

WARD: Practice time is very key. It comes down to repetition. You run those same routes four days a week—Monday, Tuesday, Wednesday, Thursday. You're going through your reads and progressions. If you've done those drills time and time again, then it becomes second nature. It's like getting in a car and driving to work. You're on autopilot.

GREER: He talks a lot about repetition. There's a long-standing theory in neuroscience by a researcher named Donald Hebb called long-term potentiation. It's useful for recognizing that with repetitive stimuli, the strength of a synapse can change.

The experiment is as follows: You have a presynaptic axon and you have a postsynaptic cell; they're links in a neurological chain. You apply repetitive stimuli to the presynaptic axon, which makes the postsynaptic cell fire. Then you wait a few minutes, and you go back and you do it again. What you find is that to get an equal response from the postsynaptic cell, the magnitude of the stimulus to the presynaptic axon can now be lower. The strength of the synapse has increased as a result of the prior experience. Or to put it another way, experience can change the way the nervous system responds.

While the mechanism of how it works may be a little technical, Hebb's concept of long-term potentiation helps us see an athlete's job in a very different light.

Ward looks back fondly on all the time he spent at practice during his career, putting in tens of thousands of repetitions, improving with each rep. Players and coaches regard this, almost dismissively, as a purely physical activity. But brain researchers like Hebb and Greer understand that something profound is also at work: Those repetitions were not only building Ward's body, they were building his brain as well.

———

Question: Paper sells for twenty-one cents a pad. What would four pads cost?

If you said eighty-four cents, you might just be on your way to being an NFL quarterback. Or not.

This question came from the Wonderlic Cognitive Ability Test. The Wonderlic is a short test similar in construction to an IQ test and contains fifty questions, some of them, like this one, so easy they seem like trick questions—only without a trick. Others are somewhat more challenging.

The Wonderlic is used in pre-employment screening for a wide variety of jobs. One of those jobs is NFL quarterback. Introduced into the draft process by Paul Brown of the Cleveland Browns in the mid-1960s, and later championed by Tom Landry, coach of the Dallas Cowboys, the Wonderlic is just one part of a battery of tests taken by most NFL players at the annual pre-draft Combine. At the Combine, prospects run through a variety of drills, ranging from 225-pound bench presses to measured vertical leaps. The physical tests seem to do a reasonably good job at parsing the differences between, say, fast receivers and very fast ones, between strong defensive linemen and ones who need to put in a little more time in the weight room.

The Wonderlic? Not so much. While test scores aren't made public, the leaked numbers that have been widely reported reveal that a large number of Hall of Fame quarterbacks put up less than stellar Wonderlic scores, including Peyton Manning's 28 out of 50, Brett Favre's 22, Terry Bradshaw's 16, and Dan Marino's 15.[*]

For all its flaws, NFL teams continue to use the Wonderlic— and a similar mental aptitude test that supplemented it in 2013.

---

[*] The all-time Wonderlic leader is Pat McInally. The Harvard grad–turned–Bengals punter is reportedly the only player to score a perfect 50 on the test. Former Giants GM George Young told McInally that his high score "may have cost you a few rounds in the draft because we don't like extremes. We don't want them too dumb, and we sure as hell don't want them too smart."

At some level, the football establishment realizes that, to paraphrase Yogi Berra, 90 percent of the game is half mental.

"I wanted to make sure Grandma could cross the street without getting killed," Jocelyn Faubert admits, but his research at the University of Montreal may enhance the lives not only of senior citizens but of NFL general managers as well.

A neuroscientist, Faubert explains that one of the problems that the elderly encounter as they age is a diminished ability to process information in complex situations. In some cases—walking in a crowded store—this can be annoying. In others—crossing the street at a busy intersection—it can be outright dangerous.

"There's a lot of information coming in, and older people are detecting the information, but they're not necessarily interpreting it well," he explains. "They slow down. They hesitate. It becomes scary. And when that happens people don't want to leave the apartment."

It ultimately became apparent to Faubert that the dangers that vex an elderly pedestrian on a busy street corner are not much different from those faced by a quarterback in the pocket. It's all information processing.

A decade ago, Faubert got funding for a device that would help him study complex information processing in minds both fast and slow. Until that point, most brain researchers were interested in easily controlled, small-scale experiments on desktop screens. In an attempt to replicate real-world conditions more closely, Faubert created immersive environments—known colloquially as "caves"—with 3-D images projected in a room-sized booth.

"I wanted quite a large visual field. And there had to be some speed components, because we're confronted with fast-moving elements and we have to make rapid decisions," Faubert explains. "We found that we could train older observers to improve, and there was transfer of that training on socially relevant stimuli. We found that they were tremendously improved in anticipating body movements, so they can avoid collisions."

In gathering data for the project, Faubert also enlisted younger volunteers, and the results for the athletes in that group surprised him. The researchers assumed that young people with an athletic background would start much higher than the seniors but wouldn't have much room for improvement. "We found that wasn't the case," Faubert explains. "What we found is that athletes not only started off higher, but they also improved faster. There's something special about how they can process this input and especially how they can *learn* it."

Faubert soon realized that Grandma and a rookie quarterback both encounter the same information-processing challenges, just at vastly different levels. And that the same diagnostic tests that can assess Granny's ability to walk across a four-lane intersection can also predict the fledgling QB's ability to find an open receiver while avoiding a blitzing safety.

Faubert's research culminated in a device called Neurotracker. It's a commercial version of his research device which comes in several sizes, ranging from a room-sized installation to a portable version that runs through a 3D TV. While it's even been used by hedge fund traders looking to sharpen their ability to make quick decisions, Neurotracker's most interesting application has been to evaluate and train elite athletes. Among those using the device are Canadian women's hockey team goalie Shannon Szabados, mogul skiing gold medalist Jennifer Heil, and players on the University of Tennessee and the University of Oregon football teams. At least one elite NFL team uses Neurotracker at the NFL Combine to evaluate its draft prospects.

What's it like training with the Neurotracker?

In a word, challenging.

The goal is straightforward enough. The test begins with seven stationary yellow balls on a black background on the Panasonic 3D flatscreen. Four of the balls turn red for a moment and then turn back to yellow. The object is to keep track of those four previously red balls among the seven yellow balls. The balls move randomly on the screen, bouncing off the edges of the frame as well as one another, and then stop. At the end of a trial the subject now must identify the four "red" balls.

A trial lasts only eight seconds, but it's very intense. Faubert's description of the necessary skill—"hyperconcentration"—is quite apt. The task of finding those four red ones may look easy, but it's anything but. It's not terribly difficult to keep track of one or two of the balls, but when the balls are moving rapidly, it can be fiendishly hard to locate all of them.

The program doesn't test a player's reactions per se, like a video game such as Call of Duty. Neurotracker only tests the player's ability to process information. Hand-to-eye coordination isn't a factor. Neurotracker doesn't give extra credit for clicking on the appropriate balls quickly. You can even call out the numbers and have a trainer enter them. It's not about reflexes; it's about accurate information processing.

Athletes begin with the core game, in which four balls move at a constant speed throughout each trial. The speed of the balls in the next trial is keyed to the player's performance in the previous trial. So like a good coach, the software keeps upping the ante without approaching the point of overload. Once a player has established a baseline, the program starts throwing out variations. In some games, the balls speed up and slow down randomly. In others, the goal is to mark the balls in order, prioritizing the targets.

Neurotracker is all about progress. A normal person might start out at a baseline of Level 0.7 and in an hour improve by more than 40 percent to Level 1.0. But an elite athlete might start at Level 2.0, where the balls are moving at hyperspeed and bouncing around more dynamically. At those levels, a normal person can have trouble tracking even a single ball for the full eight seconds. If the four balls were actually blitzing linebackers, that failure to process information would result in a trip to the hospital.

A program like Neurotracker can provide valuable answers to an NFL GM. It can measure a quarterback's ability to process information in a complex environment. And if the player runs a series of trials, it can assess something equally important: his ability to improve his information-processing skill. All of which could be vital information on Draft Day.

For researchers, the data is rich but remains somewhat difficult to parse.

"Is this a nature-or-nurture problem?" Faubert wonders. "It's an open question still, and I think both are involved for sure. If you expose the kids very early, they build specialized circuits, and they're off the blocks faster. But if you do this to two or three or four kids, you're going to find some of them learn faster, and they're the ones who get more experience from the same stimulation, and they become experts faster. I don't think they're separable, and in the end you need both."

At least if you want to become an NFL quarterback.

PART FOUR

THE FUTURE

"I have seen the future and it is much like the present . . . only longer."

—DAN QUISENBERRY

CHAPTER 14

OF RISK, INNOVATION, AND COACHES WHO BEHAVE LIKE MONKEYS

It was early in the 2008 season, and the Miami Dolphins were already on the ropes.

They had lost their first two games, the second one an especially lopsided defeat to the Arizona Cardinals. Up next? A New England Patriots team that was riding an NFL-record regular-season winning streak of twenty-one games.

The Dolphins? They had been the worst team in football the previous season, with a 1–15 record. Miami had lost eight of their last eleven games against its division rival New England dating back to 2002.

Which is to say, this matchup did not favor the Dolphins.

On the long flight home after that disheartening loss to the Cardinals, Miami coach Tony Sparano pondered his options. He called quarterback coach David Lee up to the front of the plane.

What did they come up with? The idea of using a formation

called the Wildcat, which Lee had employed while running the offense at the University of Arkansas. In the Wildcat, Arkansas star running back Darren McFadden would take the snap at center in lieu of the quarterback. The other running back, Felix Jones, would go in motion from left to right. The quarterback, meanwhile, would be lined up at wide receiver.

From this esoteric formation came three simple plays: Steeler, Power, and Counter. In Steeler, McFadden takes the snap, hands it off directly to Jones as he streaks in front of him, trying to outrun the defensive pursuit. In Power, McFadden gets the snap, *fakes* the handoff to Jones, and then runs straight ahead through a hole on the right side opened up by his blockers. Often McFadden would decide at the last moment whether to keep the ball or hand off, depending on the reaction of the defense. In Counter, McFadden takes the snap again, fakes to Jones, waits for the defense to commit to the right side, and then he runs left. Those three plays, along with a couple of basic passes by McFadden off that formation, made up the Wildcat.

The formation seemed radical, but it was hardly new. It borrowed elements of the single wing formation invented by Pop Warner, who ran it for the legendary Jim Thorpe. In 1907.

Sparano considered the way the Wildcat had made McFadden a star—he finished second in the Heisman Trophy voting twice. He considered that Miami's two best players were running backs Ronnie Brown and Ricky Williams. And he considered his lack of other options.

"This is something really interesting we could do in a ball game," Sparano told *The Washington Post,* "if we had enough nerve to bring it out there and call it."

For an NFL coach like Tony Sparano, what's the goal in each game?
  To win?
  Not so fast.
  Sparano, like any other NFL coach, will tell reporters or anyone

else who'll listen that his job is to win the game. But deep down he has a meta-goal that takes precedence over winning: to keep his job so that he can coach the *next* game. Usually, winning the game at hand is a pretty good way to accomplish that. But not always. Sometimes it might be better to lose while following the conventional wisdom than to win by employing outside-the-box strategies that would encourage fans, the media, and front-office personnel to second-guess him. Going for it on 4th and 2, benching the starting quarterback, using an offense that hasn't been employed in the league in fifty years—those are all ways for a coach to attract unwanted attention.

Most pro football coaches—who might not admit it—make decisions with this dynamic in mind, playing the game differently when they're being scrutinized than they would in a vacuum.

Werner Heisenberg would understand. The German physicist won the Nobel Prize in 1932 for the discovery of quantum physics. The pioneer of the Uncertainty Principle, he observed a similar phenomenon in the physical world. Heisenberg argued that the mere process of observing an experiment can and often does change its outcome. For example, dipping a thermometer into a liquid to measure its temperature actually changes the temperature of the liquid slightly. A tire gauge lets out a tiny puff of air as it checks the pressure of your steel-belted radials, lowering the psi a small, but real, amount. "What we observe is not nature itself," Heisenberg argued, "but nature exposed to our method of questioning."

He called this the Observer Effect.

And on some level it's clearly at work in the NFL. Which raises a question: What is it that spurs innovation in the NFL? And perhaps even more important, what *prevents* coaches and teams from innovating?

With nothing to lose, the NFL's worst team prepared to face the best team, and Sparano's Dolphins ran the Wildcat in practice. Some of

Miami's veterans were skeptical, noting that the unconventional formation forced players out of their comfort zones.

But on Sunday, against the Patriots, the Wildcat was successful beyond anyone's wildest dreams. Miami ran six plays out of the formation, and four of them resulted in scores. Brown ran for three TDs and threw for another. Miami won 38–13, and that lopsided score ended the Patriots' streak.

"They were like, '23 quarterback! Go over here! No, go over here!'" said Brown, describing the reaction of the normally disciplined Patriots defense. "So I was like, 'Okay, we got them on this one.'"

"We had trouble with their new stuff. We had trouble with their old stuff," said Patriots coach Bill Belichick after the game. "We didn't play very well on defense. We didn't coach very well. We didn't play very well across the board, and they did a good job, so give them credit."

"The theory that coaches were purely motivated by job security and didn't want to go against the conventional wisdom, that didn't quite satisfy me," says Brian Burke of Advanced NFL Stats. Week in and week out, Burke analyzed games and saw evidence that NFL coaches were costing their teams yardage, 1st downs, and, ultimately, games because of the questionable decisions they were making. He further noticed that those coaches almost invariably erred on the side of caution. But why?

Burke argued that there's something bigger going on when a coach punts on 4th down when an objective analysis shows he should have gone for it. Or keeps losing with the same tired passing plays when he should toss out his offense or at least replace his quarterback.

It turns out something bigger has been going on for 35 million years.

Daniel Kahneman never took an economics class, but that didn't stop him from winning the Nobel Prize in economics. Kahneman won the award in 2002 for his work in prospect theory, which helped to create the field of behavioral economics.

Before Kahneman and his longtime partner, the late Amos Tversky, came along, economics was based on one overarching assumption: that everyone in the economy would act "rationally" in an attempt to maximize their wealth. And in those instances when someone didn't aim for maximum profits, economists assumed that they simply lacked the information. They would aim for a maximum return if they *could,* but they can't, so they don't.

Kahneman didn't buy it. He observed that human beings consistently acted irrationally when it came to making decisions about money. But what made Kahneman's work compelling—and useful— is that while these behaviors were irrational, in economic terms, the *way* in which people deviated from an ideal strategy was predictable.

Prospect theory argues that economic decision-making is, like Einsteinian physics, *relative.* The main tenet of this school of thought is that humans make economic decisions not in absolute terms, in the way a computer might, but relative to personal reference points, and they make a series of predictable errors because of that. And emotions definitely enter into it. "You want to get someone pissed off? Just tell them that everyone they work with now makes a dollar more than they do," explains Yale behavioral scientist Laurie Santos. A computer sees a one dollar raise as insignificant. A guy in a cubicle most certainly does not.

One of the main principles of prospect theory is that humans tend to treat losses very differently than gains. Researchers have discovered that humans are consistently about two and a half times more sensitive to losses than gains. For example, if you've ever had a day when you found a twenty-dollar bill on the street and smiled about your good fortune, then later found a twenty-dollar parking ticket on your car and were ready to declare World War III because of this disaster, you understand this on a personal level. "Losses seem to hurt more than a win feels good," explains Keith Chen, a Yale economist who has worked with Santos.

This bias—loss aversion—plays out in any number of different ways in the real world. For example, stock investors repeatedly defy logic by selling the winners in their portfolios when they should instead dump the losers. A near-pathological avoidance of loss as homeowners reacted to falling real estate prices helped fuel the subprime mortgage crisis.

One of the most powerful examples of loss avoidance comes from the world of sports. Professional golfers on the PGA Tour, it turned out, were throwing away strokes—and ultimately money—by treating a putt for par differently than a putt for a birdie. A study by two University of Pennsylvania researchers analyzed 2.5 million putts and revealed that professional golfers are 3.6 percent more likely to make a putt for par—that is, avoid a loss relative to par—than they were to drain an identical putt for birdie—that is, post a gain relative to par. In general, golfers tend to leave those birdie putts short of the hole, playing them less aggressively compared to the attempts at par.

Tiger Woods explained this mindset in a 2007 interview. "Any time you make big par putts, I think it's more important to make those than birdie putts," he said. "You don't ever want to drop a shot. The psychological difference between dropping a shot and making a birdie, I just think it's bigger to make a par putt."

But, of course, professional golfers aren't playing against par. They're playing against their rivals, and saving a stroke while attempting a birdie is worth exactly the same amount as a stroke saved while scrambling for par. Yet the world's best golfers playing for huge purses treat those two situations—one for a perceived loss and another for a potential gain—very differently.

This is the way humans approach wins and losses across a wide variety of disciplines. Including professional football.

———

The Wildcat worked brilliantly for the Dolphins. For a while.

In the 2008 season, the Dolphins ran eighty-four plays out of the Wildcat, averaging 5.7 yards per play. They finished 11–5 and shocked fans and pundits alike by making the playoffs. Sparano's experiment started a trend in the league, with other NFL teams developing their own version of the Wildcat, acquiring running quarterbacks and running backs with ball-handling skills to run the retro-style offense.

By the 2009 season, however, it became clear that much of the Dolphins' success the previous year was due to a first adopter's advantage. The Wildcat offense confused the Patriots in that first meeting because they hadn't prepared for it. Why would they have? While a few college teams had experimented with the formation over the years, no modern NFL team had ever used it. Once the Dolphins started to use the Wildcat regularly and other teams jumped on the bandwagon, the element of surprise waned rapidly, and defenses took advantage of the simple fact that a team running a Wildcat offense essentially didn't have a quarterback.

Miami won seven games in each of the next two seasons.

But ultimately Sparano's risk was smart, even if its payoff was short-lived.

"In terms of team building, risk taking is good," Burke argues. He notes that in any given year the average NFL team enters a season with a Bayesian prior expectation of winning a Super Bowl that's 1 in 32. "You're a 31–1 underdog. You *want* to take chances," Burke explains. "It's okay to miss the playoffs and win five games instead of seven. It doesn't hurt you that much. But teams are very conservative. They'd rather win a few more games and avoid having a terrible record."

Burke cites as an example the Saints signing quarterback Drew Brees. "When the Saints picked him up as a free agent, there was a chance that he was going to be a Hall of Famer, and there was probably an equal chance that he was never going to be able to throw a ball again because of the surgery on his shoulder. That's okay. You *want* to do stuff like that. That's a perfect example of a very high-variance strategy helping a team that had very little success in its entire existence to become a Super Bowl champion in just a couple of years."

Are the coaches and general managers who are playing it close to the vest doing what they learned from their mentors, their parents, and their second grade teachers? Or is there something deeper at work? Are we *wired* this way?

To find out, Santos, a Yale behavioral scientist, and Chen, a Yale economist, taught monkeys to use money.

The first step in their ambitious program of cross-disciplinary research was to train their capuchin monkeys, a task that proved surprisingly easy. The decision to work with capuchins was, at first, a practical one. They're smaller and much less dangerous than larger primates like chimpanzees, but almost as smart.

In the wild, capuchins are quick learners and persistent problem solvers. In the tropical forests of South America, for example, they're famous for their ability to use stones of different sizes and densities as hammers and anvils to open rock-hard palm nuts, which are difficult to shell, even for humans. One of the most compelling moments of the BBC's *Earth* series shows a 20-pound capuchin picking up a rock that weighs almost as much as he does, lifting it over his head, and crashing it down sledgehammer-style to exploit a tiny weak spot on the nut's shell. The monkey drops the large rock again and again, weakening the shell with every blow, until the nut finally cracks. One part poetry in motion plus two parts dogged perseverance equals a mid-afternoon snack.

When Santos visited the capuchins at another university before she had set up her own lab, she noticed the lock on the enclosure door and assumed that it was there to keep out animal rights protesters.

"If you don't keep it locked, the capuchins will figure out a way to open any latch in a second," the researcher warned Santos.

The monkeys in Santos's lab quickly learned to trade small metal tokens for a slice of fruit. They were soon treating the tokens as something of value, as if they were pieces of food. Soon it became clear that the monkeys were sensitive to changes in price, which was the real threshold between a stupid pet trick and a sophisticated economic behavior.

Chen pushed the experiment further, introducing the concept of risk to the Monkey Market. To test their appetite for gambling, the researchers began to throw wrinkles at the monkeys, introducing "risky" sellers who would show the monkeys one quantity of food and then hand over another. Seller A would show the monkeys only one grape but then *add* a second grape before handing over two. Seller B would show the monkeys three grapes but take one back before handing over two.

It was immediately clear that the two scenarios "felt" very different to the monkeys. And to a human, for that matter. But do the math, and the end result is exactly the same. Each researcher always ends up giving the monkeys two grapes. A computer would value each of the sellers equally. And yet the monkeys greatly preferred dealing with the "generous" Seller A, who consistently adds a second grape, than the "stingy" Seller B who takes a grape away.

These results—choosing the safer route when faced with a bonus but gambling to avoid a loss—suggested that the monkeys in this New Haven lab expressed the type of loss aversion that Kahneman discovered in humans. And the monkeys' responses, in these experiments and others, mimicked those of human subjects not only in kind but in degree. Chen notes that the data set for the monkeys—which showed that monkeys would be 2.7 times more likely to make a risky decision to avoid a loss than to gain a bonus—was completely indistinguishable from what you might find in a human trial using a bunch of Yale undergrads. Or a collection of NFL coaches.

But the real takeaway is even bigger. Unlike chimps and other great apes, which are close genetic cousins to humans, capuchins branched off from our section of the evolutionary tree a very long time ago—about 35 million years. Which means that the impulses that make us treat winning and losing so differently can be traced back to the deepest parts of our behavioral coding. So when the coach of your favorite team elects to punt instead of going for it on 4th down, cut him some slack. He's only doing what he's been programmed to do.

CHAPTER 15

DESPERATION PLUS INSPIRATION EQUALS 16,632 ELIGIBLE RECEIVERS

Kurt Bryan walked in and couldn't quite believe what he saw on the whiteboard. Steve Humphries, Bryan's offensive coordinator at Piedmont High School in northern California, had diagrammed an offensive formation using not one but two quarterbacks. He wasn't sure if he should (a) assume it was a joke and laugh or (b) assume it was not a joke and wonder if Humphries was working too hard.

Where does an idea this profoundly crazy come from? A fan's frustration. Humphries was and is a devoted San Francisco 49ers fan, his passion dating back to the championship teams of the 1980s and 1990s. He especially loved the team's quarterbacks. And for a brief time, the 49ers had not one but two Hall of Fame passers on their roster. But while Joe Montana took the snaps, Steve Young sat on the sideline and watched. That seemed fundamentally wrong to Humphries.

"You had one left-hander and one right-hander. Steve Young was a great runner, while Joe Montana was more of a pure passer,"

Humphries recalls. "Why couldn't you have *both* of these guys playing at the same time? The defense wouldn't even know which one was going to get the snap."

Little did he know that his wacky impulse to double the fun of NFL football would lead to an offense with 16,632 options. Give or take.

Bryan immediately identified the problem with Humphries's dual-QB set: *It didn't go far enough*. He replaced Humphries's strange set with an even more profoundly radical formation in which *every* player on the field potentially became an eligible receiver.

The A-11 offense started out in the way so many innovations do: 1 percent inspiration, 99 percent desperation. Piedmont High School's Highlanders were a mediocre 5–5 team without real prospects for getting better anytime soon.

"We took a physical pounding," Bryan explains. "We were outmanned."

The school owned a reputation for churning out Ivy-bound seniors, not NFL prospects. Academically, they ranked seventy-third in *U.S. News and World Report*'s national high school rankings, and the winners of the school's nearly half-century-old bird-calling contest landed a spot annually on the *Late Show with David Letterman*.

But in football it was a different story. Piedmont is a relatively small school, with a student body of around 800, and their athletic teams compete against schools with student bodies almost twice as big. Whatever Piedmont boasted in SAT savants and campy California quail callers, they lacked in 300-pound athletes. Running backs? Those we got. Offensive linemen? Not so much.

So Bryan and Humphries focused on finding an innovative way to use the athletes that could be recruited from the school's soccer, lacrosse, and basketball teams. Good medium-sized athletes. Smart ones who could be trusted to learn and execute complex assignments. Like running an offense that some called radical, and others just called crazy.

Bryan pitched his idea to the principal, who not only approved the idea but also actively worked to shelter Bryan from the mountain of negative feedback his radical game plan would ultimately receive. And thus the A-11 offense—which might very well provide a preview of the future of football—was born.

Taking a page from tax accountants everywhere, the A-11 offense exploits a loophole in football's rule book. On a normal play, the five players on the offensive line are not eligible receivers, meaning they can't catch a forward pass past the line of scrimmage. Only the remaining players—the wide receivers, the tight end, the two running backs—can catch a forward pass legally.

The ineligible receivers wear numbers between 50 and 79.[*] However, on certain kicking plays, the numbering rules about eligible receivers are temporarily thrown out the window. Bryan and Humphries pored through the rule book carefully and realized that in something called the scrimmage kick formation they could leave their quarterback 7 yards deep—where the kicker would normally stand—in a kind of long-distance variant of the shotgun formation. The same quirk of the rules that allows a punter to try to throw for a 1st down on a bad snap allowed Piedmont's quarterback to set up in a way that allows everyone else on the field to remain potentially an eligible receiver. If Bryan and Humphries positioned their quarterback just so, they could bypass the jersey numbering requirement and use this special teams' variation as their basic offense. And make the defense contend with a plethora of potentially eligible receivers on the field.

In practice, the A-11 offense bears only the most tangential relationship to normal football. Players are arranged in three "pods" spread across the field along the line of scrimmage. The players on the right and left pods are wide receivers. The middle pod consists of

---

[*] A player wearing an "ineligible" number can, however, formally declare himself to be an eligible receiver to officials before the play. The player who used this strategy most successfully was Patriots linebacker Mike Vrabel, who lined up at tight end in short-yardage situations and caught ten passes in his career, all for touchdowns.

a center with a tight end on either side. All of the players are wearing jerseys that make them eligible receivers, although once they get into formation, the center, the two other players in the center pod, and the innermost player on each of the outside pods become ineligible receivers because of their place in the formation. Keep in mind, though, that ineligible receivers aren't restricted to just blocking—they can still catch laterals and run with the ball. The only thing they're prevented from doing is catching a forward pass.

"You want to create eleven islands, as many one-on-one match-ups as you can, and get the ball to the best player," Bryan explains.

The defense doesn't know which five players will be ineligible on any given play until the players take their positions at the line of scrimmage. The A-11 is run out of a no-huddle, which gives defenders less time to react to the personnel changes. As you can imagine, this creates matchup nightmares for the defense. When the play begins, the center makes the long snap to one of the two quarterbacks positioned 7 yards behind the line.

And then the fun begins. At least if you're running the offense.

What would an NFL play look like with more than one quarterback? Rewind to a Carolina Panthers game on December 4, 2011.

It wasn't much of a play at the time. No one scored a touchdown. No one broke a record. And the game itself was neither particularly close nor especially important. But if you look closely, you can see in that play one possible future for pro football.

Most of the way through the 2011 season, the Carolina Panthers were a bad team getting better. The Tampa Bay Buccaneers were a bad team getting worse.

In the middle of the first quarter, the Panthers jumped out to an early 7–0 lead. Now they have the ball again in Tampa territory. It's 1st and 10, and the Tampa Bay defense is back on its heels. Just moments earlier the Bucs were feeling good about themselves—they had stuffed the Panthers on two consecutive short run plays to force a 4th down. But their hopes are dashed when Carolina's flashy young quarterback Cam Newton takes over. He fakes a handoff to the right, and while the defense follows the play-action misdirection, the rookie from Auburn instead bootlegs left to collect the 1st down easily.

Newton next lines up in the shotgun, 3 yards behind the center. He takes the snap, but instead of looking downfield and reading the defensive coverage on his receivers, Newton smoothly pivots to his right. In lieu of firing the ball downfield, he throws the ball *backward* to his wide receiver Legedu Naanee, who is a yard or two farther behind the line of scrimmage. Once upon a time, they called this a lateral.

When Newton releases the ball, four defensive linemen and three linebackers cut left to frantically pursue Naanee. But before any of the Tampa defenders can get within a couple of yards of the receiver, Naanee casually tosses the ball *back* across the field to Newton. The quarterback doesn't have a Tampa defender within 10 yards of him when he makes the routine catch, and but for Tampa's defense scurrying to and fro, the throw seems as easy as if Naanee were tossing the ball back to Newton during warmups. This is only Naanee's second pass in the NFL, the other being a 21-yard completion to running back LaDainian Tomlinson when he was playing for the San Diego Chargers in 2009.

Newton catches the ball at the Tampa 35-yard line and tucks the ball under his left arm and heads downfield. Naanee's throw to Newton is just barely a forward pass, but if it had been another lateral, Newton would have had the option of running, throwing a forward pass, or even another lateral. On this play, however, his only option is to run downfield. Newton possesses 4.4 speed, but now he is merely jogging. There isn't an orange jersey in sight, but Newton also realizes that he's about to overrun his own blockers, who were a little slow to react on this unusual play, although much less so than Tampa's utterly bewildered defense. It isn't until he reaches the 20-yard line that the rookie begins to showcase his dazzling speed.

No one even touches Newton until he's inside the Tampa Bay 10.

The traffic jam among Newton's blockers has given Tampa's defensive secondary a moment to recover, and veteran defensive back Ronde Barber makes a desperate shoestring tackle at the 2-yard line to save the touchdown. Three plays later, the Panthers score anyway.

But that was a mere afterthought.

What was stunning was the profoundly casual nature of this play. Most NFL plays are executed with a well-drilled, almost military precision. Not this one. It looked like the sort of improvisation that you might find in a backyard pickup game, something conceived on the spot and diagrammed in the dirt, with assignments explained by telling someone's brother-in-law: "You fake going deep, and then run back toward the tree. I'll throw the ball to you. You hold the ball, count to three-Mississippi and toss it back to me." It also looked a little like a play from an NBA basketball game, where the pass is a means to an end rather than an end in itself—Tony Parker, say, of the San Antonio Spurs tossing the ball to Manu Ginóbili, who, of course, has the option of taking the shot or just making another pass.

Naanee, for his part, wasn't a quarterback, although he was recruited as one in college at Boise State. He does have skills, though. When he was with the Chargers, he was the passer on the scout team, serving as the stand-in for Titans star Vince Young in preparation for a playoff win. In any case, Naanee had more than enough quarterbacking chops to make this fascinating play work.

But it's Newton who was the wild card in this play. The Auburn product was the first player chosen in the first round of the 2011 NFL draft. At 6'5", 245 pounds, with world-class speed, he's an almost otherworldly athletic specimen. What really makes him special—and a possible harbinger of things to come—is the brain that runs that body. He has the vision to run and pass with almost equal effectiveness on any given play. He modeled his game on that of trend-setting quarterback Michael Vick, but Newton readily admits that he aims to pass *and* run better than Vick ever did.

By adding a second quarterback to this otherwise routine play, the Panthers did much more than double their offensive options. When Naanee tossed the ball back to Newton, the probabilities increased at a dizzying rate—and this play only began to explore them.

It's easy to dismiss this diamond-in-the-rough innovation as a trick play. Analyst and former head coach Jim Mora called it just that during the FOX Sports broadcast. Then again, sportswriters back in 1919 dismissed it as a fluke when Babe Ruth, then still a pitcher, would hit a home run, writing it off as an outlier in an era when runs were scored by linking bunt singles and stolen bases. It's not always easy to see the future, even when it's unfolding before your very eyes.

To get a real sense of just how disruptive an offense can be when everyone's an eligible receiver, you need to crunch the numbers.

A conventional play has thirty-six possible options. Six of the eleven players are eligible receivers, and each of them can pass the ball to any of the other six eligibles. Six times six equals thirty-six.

In the A-11, the numbers are vastly different. To begin with, any one of the eleven players can either be one of the five linemen or one of the six eligible receivers. So before the play starts, you have to decide who the eligible receivers are. The math begins to get a little involved here, but basically there are 462 unique ways in which you can break the team up into five linemen and six skill players. Then, for each of those 462 possible arrangements, you once again have the standard thirty-six possible outcomes as far as who takes the snap and who ends up with the ball: $462 \times 36 = 16{,}632$.

And that's 16,632 possibilities *before* running the play.[*]

---

[*] If you choose your six skill players one by one from your eleven players on the field, you have 11 choices for skill player 1, 10 choices for skill player 2 (since one player has already been picked), 9 choices for player 3, 8 choices for player 4, 7 choices for player 5, and 6 choices for player 6. That's 332,640 possible permutations ($11 \times 10 \times 9 \times 8 \times 7 \times 6$). But since the players are functionally identical, the order doesn't matter. So you divide the 332,640 by the number of times each group of players repeats as an identical group with different ordering. That's 720, and 332,640 divided by 720 equals 462. (See Notes.)

Mathematics only takes you so far on the football field. Piedmont's first game running the A-11 was ugly. The system that was supposed to fool opposing defenses instead fooled Piedmont's own players.

"We got our asses kicked," says Bryan, recalling that he threw up outside the locker room at halftime. "We got blown out. It was too much of a quantum leap." Piedmont's second A-11 game was somewhat better from an execution standpoint, but the score was still lopsided. At that point, Piedmont's offensive line coach quit, and a player mutiny was brewing. Bryan received a letter from a parent who worked as a scientist, suggesting that after two losses the A-11 was a failed experiment.

Bryan saw things differently: The A-11 was a fundamentally sound idea with flawed—but improving—execution. Piedmont had a road-trip game coming up, and both Bryan's career and the team's A-11 experiment stood at a crossroads.

"If we went 0–3, my house was going to be sold by the time I got home, and not by me," Bryan jokes. But Piedmont would run more A-11, not less. They won the next game. And then seven more in a row.

"The 300-pound linemen couldn't handle the track meet we threw at them," Humphries explains. The scheme attracted national attention on ESPN and in publications ranging from *The New York Times* to *Scientific American*.

The David and Goliath story line had built-in appeal, but as a strategy, the A-11 featured more than that. As football continues its inexorable drive toward increasing complexity, based on the search for order hiding within apparent chaos, the A-11 seemed more and more like the shape of things to come.

There are two sides to the A-11: What the plays look like. (In a word, *weird*.) And what the players look like. (In a word, *small*.) With any of the eleven players potentially running a post pattern, most of the players on Piedmont, or on any of the hundreds of high school teams using the A-11, looked like wide receivers or halfbacks. Bryan contends that the trend would continue if the offense were run at an elite level in Division I college or even the NFL.* As positions became more specialized over the years, offensive linemen grew to twice the size of running backs. Freed of those positional responsibilities, Bryan and Humphries predict that any team running an A-11 offense would feature eleven medium-sized athletes and that the defense that had to chase them would ultimately follow suit.

So in that way the A-11 is both a peek into the future and a long look in the rearview mirror. This offense harks back to a day when the game resembled rugby and the players didn't have defined positions.

Who's the ultimate A-11 player? Bryan doesn't hesitate. "Tim Tebow."

The guy who can't quite throw well enough to be an elite quarterback but runs better than any quarterback has a right to and has the size to catch a pass over the middle and the speed to break it for a long gainer. Humphries contends that the A-11 could make use of other skilled players—like Eric Couch of Nebraska—who don't quite fit the physical mold of an NFL quarterback.

The A-11 player is a hybrid, with a variety of skills that don't easily slot into a single position. It could be football's answer to an NBA All-Star Game, where the sport's best players defy being pigeonholed into tightly defined positions. Small forward LeBron James mates a point guard's ball-handling skill with a power forward's body, power forward Dirk Nowitzki passes the ball like Bob Cousy, and 6'6" combo guard Kobe Bryant has the size to post up any point guard who dares to defend him.

The NFL may already be beginning to heed the lessons of the A-11. A number of the league's best young quarterbacks—Newton, Washington's Robert Griffin III, San Francisco's Colin Kaepernick,

---

* Which could only happen, of course, if the NFL or NCAA were to make a few crucial rule changes.

Seattle's Russell Wilson, and even Indianapolis's Andrew Luck—have a broad-based run/pass skill set that would shine in a more wide-open offense. This is in sharp contrast to the previous generation of elite quarterbacks—Peyton Manning, Tom Brady, Drew Brees—who are more classic pocket passers.

After two seasons, the powers that be in California high school football changed the rules to close the scrimmage kick loophole. Teams could still run elements of the A-11, such as using two or more quarterbacks, but the rules that allowed teams to avoid the jersey numbering requirement—and thus put eleven potential receivers on the field—bit the dust. Piedmont played two more seasons of a hybrid A-11 with modest success, before Humphries and Bryan left the school with an eye toward starting a pro league that featured A-11 rules.

While the goal of the A-11 was simply to win some high school football games, its radical nature changed things enough that it shed light on some of the problems with the pro game. For example, Bryan and Humphries tout the A-11 as a potential solution to the head injury problems facing professional football. "In two years of running a pure A-11 they didn't have a significant injury on offense," says Bryan, adding that the team continued to remain injury-free during the seasons they ran a modified A-11 after the rule changes. Of course, that's a very small sample. Playing an actual A-11 game with bigger and stronger athletes and more sophisticated schemes on both offense and defense would reveal if this revolutionary offense is a panacea in this regard or merely a conduit for different kinds of injuries.

And maybe that will happen sooner rather than later as A-11 concepts creep into the offenses of major colleges, like the University of Oregon, and into the pros, including the Colin Kaepernick–led attack that took the 49ers to Super Bowl XLVII. Talking to *ESPN The Magazine*, Pittsburgh linebacker James Farrior explained that the real future of the A-11 lies with some undersized but athletic football player who encountered this crazy offense in high school and then enrolled in one of those Ivy League schools that provide an alternate route to running a pro football team.

"It's evolution," says Farrior. "One of those kids in that offense is gonna grow up and be an NFL coach one day, and he'll have this system in the back of his head, waiting."

CHAPTER 16

THE MAN WHO LOVED TACKLING

"Wrap him up."

Since the days of Pop Warner, coaches have used these three words as a mantra when teaching tiny football players how to tackle. And as mantras go in the often-martial world of football, this one seems benign enough, conjuring up images not of broken bones but of blankets and burritos.

But when Bobby Hosea hears those three words, he cringes. The way he sees it, those words are at the very center of football's injury problem. Hosea is football's accidental evangelist, a former-player-turned-actor-turned-amateur-biophysicist who's made it his life's work to change the way that football players tackle one another. His mission started simply enough, as a way to try and save his preteen son from getting hurt. But before it's all said and done, he may end up saving football.

———

"When I say 'wrap him up,'" Hosea challenges, "what do you think of in your mind's eye?"

In explaining the problems with football's tackling truisms, he resorts to examples. It's the teacher in him. He embellishes the scenario. Imagine you're a linebacker and Barry Sanders is sprinting toward you. You're the last thing, the *only* thing, between the Hall of Fame running back and the goal line, so tackling him *really* matters. Get ready. You're about to take aim at Sanders's torso and envelop him in a big ol' bear hug. But Hosea interrupts the imaginary hit with a question.

"What's *happening* in the hug?" Hosea asks in good Socratic fashion.

The answer conjures up a prototypically athletic stance, arms in front, leaning forward, shoulders down, hips back.

"Where's your head?" asks Hosea.

It's down.

In this classic tackling stance, the crown of the helmet is leading the way. Even if you hit with the shoulder pads only, the noggin is still out in front and very vulnerable.

And that's where "wrap him up" and other truisms like, "bite the football," become dangerous.

"What the coach really wants you to do is put your head on the football," says Hosea. "The model that everybody's using from day one is broken. It can't be fixed, and it needs to be replaced.

"What I came up with was science," he continues. "I didn't know it was science, but it is."

Sixteen years ago, Bobby Hosea got the question that he hoped would never come. His twelve-year-old son asked him if he could play tackle football.

"I should have gotten him a five-iron instead of that rubber football," Hosea jokes now.

Hosea understood the dangers of football firsthand. He had

played defensive back at UCLA and had played as a pro for three seasons in the Canadian Football League and two in the USFL before moving on to a career as a character actor, landing parts in everything from blockbusters like *Independence Day* to TV movies like *The O. J. Simpson Story* to popular series like *CSI: Crime Scene Investigation.*

Since he really couldn't, in good conscience, tell his son he couldn't play, Hosea went to see the commissioner of the local Pop Warner league to ask about the league's safety policies. The commish, who was a UCLA fan, remembered Hosea from his playing days.

The commissioner said, "Hey, you're the head coach."

After the first practice, Hosea's new predicament snapped into focus: "When this started, I had one little kid, and I was scared of him getting hurt. Now I had twenty-five kids, and I was just as scared about one of *them* getting hurt."

What did Hosea do?

"I said a prayer. I slept on it. And when I woke up, something told me to go to the Building Emporium. I asked the guy for those pipes they use in sprinkler systems."

"PVC?" the guy asked.

"I guess," Hosea replied.

Hosea left with some pipe, a few tools, and an idea.

And working with his own twelve-year-old son and his friends as guinea pigs, Hosea used gardening supplies to reinvent the art and science of tackling. Instead of aiming a shoulder at Barry Sanders, Hosea's players drive with their hips. They explode upward with their torsos and arms, while actively arching their backs and pulling their heads up and away from danger. The system is based on simple biomechanics, that the hip bone is connected to the head bone by way of the backbone.

"The first lesson you learn is, 'I'm in control of where my head moves, and it moves up and away because of how I moved my hips,'" Hosea explains.

But Hosea's a showman too, so his kids get the juiced-up version, describing this new school of tackling with the kind of over-the-top language that kids eat up. "You turn yourself into a surface-to-air missile," he explains. Or "I'm just going to dissect your momentum."

Hosea's sneakily named B.I.G. H.I.T.T.S. curriculum begins by having his players tackle a big pile of nothing. "The first thing we hit is air," Hosea explains. The young defenders build muscle memory by hitting imaginary runners and landing on mats. Later they move on to foam bags. It takes several sessions before they even *consider* hitting another player. The end result is a brand of tackling that's based on leverage—Hosea admits there's some truth to the old adage "Low pads win, high pads lose"—but he is most concerned about keeping the head out of harm's way.

"It's almost like you're trying to do a backward somersault," he continues, explaining a drill called the Shamu. Hosea has devised many more drills, dozens of them, but they all have a common goal of moving the head *away* from the point of impact.

"If you drop your chin, you're doing it wrong," he concludes.

"Every year there are 489,000 injuries in youth football, at the cost of $6.89 billion," says Hosea. "If this were happening to puppies . . ." His voice trails off, the salesman of safe tackling briefly left speechless. Hosea cites statistics that suggest that fully 85 percent of head injuries happen in practice. He then notes that the knee-jerk response by youth football officials has been to simply cut back on practices.

"We're not getting them to practice *better*. We're just getting them to practice *less*," he says with exasperation.

When a high school or college player somewhere in the United States suffers a catastrophic injury—a spinal injury or worse—Hosea immediately contacts the school and offers his services to teach his techniques to the rest of the players to prevent this from happening again. His offer is invariably declined. Why? While no one will say so, Hosea believes that it's because the school's lawyers explain to the administrators that a jury would view a sudden change in teaching techniques as a tacit admission of guilt. And that could cost the school millions of dollars in court.

"I'm a big football fan, but I have to tell you if I had a son, I'd have to think long and hard before I let him play football." This comment made news, because that father of two daughters also happens to be the nation's Sports Fan in Chief, President Barack Obama. In expressing his concern about letting kids play the game, Obama joined unlikely figures like Tom Brady's father and Hall of Famer Harry Carson. While these figures are expressing natural parental concern, concussion and head injury researchers are discovering that young people are at a special risk for head injuries that goes beyond that encountered by older players.

"I would keep them out of youth football with a passion," says Robert Cantu, a clinical professor of neurosurgery at Boston University and one of the nation's foremost experts on head injuries. His biggest concern is that children's brains are still growing in ways that make them more susceptible to head injuries. "Youth brains are not fully myelinated," he explains. Myelin is a layer around the axon, which is the part of the nerve cell that emits the electrical signal. Functioning a lot like the insulation on a wire, myelin has two roles. It helps the brain's circuits conduct electricity, and it provides physical support to structures in the brain.

"If you did not have the coating on the telephone wire, it would be easier to pull it apart. The same is true of the myelin in the brain," Cantu explains. "It gives the fibers better strength, so they are better able to sustain acceleration forces that happen when you crack heads." Less myelin means less resistance to concussive forces.

And when a concussion does happen, the growing brains of young people also take longer to recover, according to Cantu. They're more prone to cytotoxic shock, the metabolic disruption in the nerve cell that happens as a result of a concussion, and they recover more slowly than an adult.

The concussion risk stems from the hidden vulnerabilities not only of young brains but of young bodies as well. Neck muscles protect players of all ages from head injuries by keeping the head from sudden accelerations and decelerations. Children have relatively weak necks, coupled with disproportionally large heads. By the age of five, for example, a boy's head has grown to 90 percent of his adult size. Simple physics, for instance $F = ma$, illustrates how a more massive head on a weaker neck means more dangerous acceleration for the young brain. In an older person, the strong neck muscles join the mass of the head to the mass of the body, which slows acceleration and thus potential damage to the brain. In a young person, the neck is more like a Slinky, allowing the head alone to accelerate faster and leaving the brain subject to greater damage. "That bobblehead doll effect means that brains are going to be rattled much more than an adult with a strong neck," says Cantu.

On a practical level, youth leagues also tend to play with older equipment, no medical staff, and amateur coaches, who aren't always familiar with the symptoms of a head injury. Cantu's final caution isn't a medical one but a philosophical one. Adult football players understand the risks they're taking; young players do not. Cantu maintains that young people are not capable of providing informed consent. "They really don't know what they are subjecting themselves to," he says.

Hosea's technique is radically different from anything taught by conventional football, and its contrarian nature has proved to be an obstacle. But the one thing it has going for it can be summed up in two words: It works.

One of the graduates from Hosea's very first class of Pee Wee players is Dashon Goldson, and Hosea couldn't have asked for a better honor student. Goldson is not only a professional football player, he's an All-Pro; he starred on a 49ers team that made the Super Bowl, a game in which he forced a fumble using the tackling technique Hosea taught him. After the 2012 season, he signed a five-year,

$41.25 million contract—with at least $18 million guaranteed—with the Tampa Bay Buccaneers. Browse YouTube and you can find a collection of Goldson's Greatest Hits:

Dashon Goldson Wrecks Mike Williams
Dashon Goldson lays down the wood vs. Browns
And simply: The Biggest hit of the 2012 NFL Season

If you know what to look for as you watch those videos, you can find Hosea's principles at work on the highest levels. Goldson explodes upward into the ball carrier while moving his head out of the path of the impact.

If you're not attuned to these finer points of tackling technique, you just see a defensive back flattening ball carriers. At the same time, because Goldson is tackling with his torso and not his head, the ball carriers aren't getting hit in the head either.

It's easy to see a glimpse of a possible future for the NFL in these clips, one in which players continue to hit hard but at far less risk to their heads.

"At every level, his coaches and the other players would ask, 'How do you tackle like that?'" Hosea reports.

The physics of Hosea's techniques also stand up to quantitative testing. In 2011, Hosea recalls that researchers hooked a helmet with sensors and had youth players hit a tackling dummy using Hosea's methods, followed by a "wrap 'em up" tackle. Under these controlled conditions, the players using Hosea's methods registered only 43 percent of the impact of a more conventional hit. And in a real game, where the opposing player is hitting a moving target and is more likely to make accidental contact with the crown of the helmet, the actual reduction is likely to be even larger.

"We want it to be the standard of care," Hosea explains.

Coaches have long known that football is a game of hearts and minds.

Hosea understands the effectiveness of that venerable model, and he's more than willing to turn it to his own devices.

Before he can teach his tackling method, he needs his players to buy in. With younger players, he jokes around, handing a kid his car keys and telling him to go drive to get donuts. "Oh, you don't know how to drive? Well, you don't know how to tackle either."

When he's talking to an older player at one of his clinics, a teenager who's already hit and been hit, one who knows how to "wrap 'em up" and has done it dozens of times, Bobby Hosea knows that he needs to break through his protective shell.

"Why are you here? Why aren't you still in bed on a Saturday morning or at home watching cartoons?"

The player shrugs.

"What scares you to death about being on the football field?"

"Nothing."

"What's your mother scared of?"

Bobby Hosea has pushed a button. The kid's mother has a laundry list of injury-related worries that has brought this young player to Hosea's clinic. Hosea doesn't need to get into specifics.

"How do those things happen?"

"You put your head down."

"How does that happen?"

"I don't know."

"How do you keep your head up?"

"I don't know."

Then Bobby Hosea tells the story of Al Lucas. It starts out like an inspirational tale. The son of two local politicians in Macon, Georgia, Lucas was a legend, a nimble 300-pounder who made everyone stop and watch. "Going against him was like going against a brick wall," said teammate Robert Vanzant. At Northeast High School in Macon, Georgia, Lucas set a school record by bench-pressing 440 pounds.

But there was more to Big Luke than just size and strength.

"That huge body was extremely intimidating, but he was so gentle," said Raynette Evans, the athletics director in Bibb County. "On the field he was as aggressive as you want in a player, but off the field he was what you would call a gentle giant."

Lucas went on to Troy State, where he earned a degree in criminal justice and won the Buck Buchanan Award as the nation's best defensive player in NCAA Division I-AA. But NFL scouts were more concerned about a gimpy leg than they were impressed by Lucas's big body, and he was passed over in the NFL draft.

The Draft Day snub annoyed Lucas but also hardened his resolve. He tried out for the Carolina Panthers and bucked the odds by cracking the roster as a free agent. One day a skinny wide receiver walked into the dining hall, and Big Luke shouted playfully, "Get that man something to eat." That one-liner became an unofficial slogan for the defensive line during the season. Still, in a league where big, strong, and fast are the rule rather than the exception, Lucas saw only sporadic action. He played in twenty games over two seasons, recording thirty-eight tackles, two forced fumbles, and a single sack.

At age twenty-six, the 6'1", 300-pound lineman had more football left in him. He was selected with the fourth overall pick by the Frankfurt Galaxy in the NFL Europe free-agent draft but decided to stay in the United States and play in the indoor Arena Football League. He signed with the Tampa Bay Storm, where he made the league's All-Rookie Team. The Storm won the championship. For the 2004 season, he signed a three-year deal with the Los Angeles Avengers, knowing that he was just a lucky break or two away from a return to the NFL.

The Arena Football League started play after the Super Bowl and continued through the spring, so Lucas would come back to Macon, teaching during his old high school team's football season. Strolling the halls of Northeast in designer suits and Stacy Adams shoes, Lucas was the one substitute teacher that even the rowdiest class clown wouldn't mess with. As an assistant football coach, he would get players' attention by loading up the weight bar until it groaned under all that heavy metal and casually pump out a few 600-pound reps while the players looked on in awe. He collected old shoes and gloves from his Arena League teammates to bring back to his players. Macon is a depressed area with high unemployment and some low-level gang activity, and the mere fact that Lucas would return sent a message that was stronger than any inspirational speech.

Big Luke kept his actual message simple: "Play like it's the last play."

On April 10, 2005, in a game against the New York Dragons, Lucas was playing on special teams, covering a kickoff return midway through the first quarter. He collided with New York's Corey Johnson, a wide receiver who was two thirds his size. It was just another play among thousands, a typical pile of large muscular bodies in pursuit of a ball carrier.

Except that after the bodies were untangled, Al Lucas didn't get up.

He had suffered a severed spinal cord. Physicians spent a half hour attempting to resuscitate him. Al Lucas was officially pronounced dead at a local hospital. It was the first death from impact in a professional football game in almost forty years.

Perhaps because Arena League Football was a minor league sport, played while most football beat writers were reporting on the upcoming NFL draft, Al Lucas's death received little national attention. The few stories written about it focused on what Lucas meant to his teammates, his students, his wife, DeShonda, who still dreamt about her husband years later, and his two-year-old daughter, Mariah. Those reports glossed over the play itself, calling it a tragic accident.

And while Bobby Hosea is quick to point out that there was no malicious intent, nothing that even remotely resembled an illegal hit, he sees the play very differently. When he tells his players about the fate of Al Lucas, he describes it as the inevitable intersection of football and biomechanics.

"Al Lucas was six foot two, 315 pounds. He hit a guy who was six-one, 185 pounds," Hosea explains.

"The receiver went up about ten feet in the air. I show them the photos. So I ask them: 'The 315-pound man hits a 185-pound man on his kneecap with his head. What do you think happened?'"

"The 185-pound guy must have got hurt" comes the inevitable response from one of the kids, focusing on Corey Johnson in low-earth orbit.

"No, the 315-pound guy who got hit in the head with his kneecap died instantly on the field."

Hosea pauses to let that hard reality sink in.

"He hit him in the kneecap with the crown of his head. His head stopped immediately. The rest of his 315 pounds kept coming and crushed all the vertebrae at the base of his neck. It killed him instantly. Right there at the Staples Center, on the middle of the field.

"He was exercising his particular technique," says Bobby Hosea to the kids about to learn a safer way to tackle, "and it killed him."

CHAPTER 17

WHY WOODPECKERS DON'T GET CONCUSSIONS

On average, an Acorn Woodpecker bangs its beak into a hardwood tree at twenty times per second. That's 12,000 pecks a day.

Which raises a question:

Why don't woodpeckers get concussions?

That's what Lorna Gibson is trying to find out. A materials science and engineering professor at MIT, Gibson has studied the Acorn Woodpecker for years.

And her work has focused on that really obvious question. How can a woodpecker bang its head against a tree thousands of times a day—every day for its entire life—and not suffer traumatic brain injury?

Size matters.

Brain size, that is.

"Having a small brain makes a big difference," Gibson says. "It's a scaling phenomenon. The human brain is like 1,400 grams. The Acorn Woodpecker's is two grams." That's roughly the weight of two paper clips. It's roughly the same reason that small animals can plunge from a skyscraper and walk away unharmed while a human can suffer a fatal head injury falling from a stepladder.

Don't get the idea that the woodpecker is somehow pulling its punches. A woodpecker can withstand forces of 1,500 G without injury. In humans, an impact of 100 G can cause a concussion.

Unlike the Galapagos finches, which made outrageous adaptations to their environments, the woodpecker's physiology isn't all that different from other birds of similar size. Gibson reports that the woodpecker's skull is made of normal bone, and it isn't especially thick. It is thicker in the beak and spongy in other areas of the skull itself. The woodpecker does have an extraordinarily long tongue that wraps inside the skull—actually, behind the bird's eyeballs—but there are no indications that the tongue protects the brain from impact; it merely helps the bird reach the insects that pecking accesses. If they needed to, other small birds could also take up pecking without suffering concussions.

The woodpecker's head does have a protective feature. Its brain occupies roughly the same amount of space in the skull as a human brain—a hemisphere. But in the woodpecker the brain is rotated 90 degrees so that the long axis of the brain is vertical rather than horizontal.

That roughly doubles the surface area of the brain that takes the impact of the hit. The other advantage? "Their brains and their skulls don't have a lot of cerebrospinal fluid," Gibson says. "The brain is wrapped pretty tightly in the skull."

The next variable is the duration of the impact. And in much the same way that a small animal like a mouse is light on its feet, a woodpecker is light on its beak. "The duration of the impact is something like a millisecond," says Gibson. By contrast, the impacts that football players encounter are more like 100 milliseconds—10 times longer.

What does all this tell us about human head injuries? It tells us that our big brains are a double-edged sword. Our big brains allow us to recognize a problem and posit solutions. Our big brains are also the reason that football has a concussion problem in the first place.

"I just killed three of your players," said Bill Simpson to Tom Moore, offensive coordinator for the Indianapolis Colts.

At age sixty-eight, Simpson had never been to an NFL game until his friend Moore invited him to a Colts game. After the game Simpson and Moore were drinking a beer in the parking lot used by the coaches and players when Simpson asked an innocent question.

"Say, what happened there?"

"What do you mean?" Moore replied.

"Those guys being carried off on stretchers."

"That happens at every game," said Moore. "They have concussions all the time."

"Whoa!" Simpson replied. "That isn't so good."

Simpson was not just any interested spectator. His Simpson helmets were world-famous, protecting race car drivers and motorcyclists the world over. An engineer by trade, Simpson knew as much about protecting people from high-impact head injuries as anyone on the planet. He was an outsider, but an eminently qualified one.

To satisfy Simpson's curiosity, Moore sent over three different helmet models from three different manufacturers, and Simpson began by putting them through the same high-speed Snell Foundation tests that his motorsports helmets pass routinely. The football helmets didn't fare nearly as well.

Simpson brought the helmets and printouts of the tests to Moore later that afternoon.

"What's that?" Moore asked, pointing to a huge 800-G spike on the graph.

Which was when Simpson responded, "I just killed three of your players." Simpson got a couple more helmets and recalibrated his testing equipment for football's more modest import. "Then I went to my composite shop and made a shell," he recalls. The results of the drop test were promising. The other helmets passed impacts of 100 G, and Simpson's first prototype registered only 60 G.

"I'm going in the right direction," Simpson said to himself.

So after fifty-five years, Simpson sold his legendary motorsports company lock, stock, and barrel. The only thing he kept was access to the testing lab until he constructed a new facility.

"I started in earnest," he explains, "to do this thing."

One of the key differences between Simpson's helmets and those being used in the NFL was his use of advanced materials. Modern-day football helmets are still made of polycarbonate, a material similar to what was used in the football helmets of the 1950s and 1960s. "Polycarbonate was banned from auto racing in 1970," he says. "We've taken some of those polycarbonate shells and put our liner material in it, and the numbers don't change. It doesn't spread a load."

Simpson made his prototypes the same way as his motorsports helmets, from carbon fiber and other high-tech composites. It wasn't easy, and it wasn't cheap. "A polycarbonate helmet could be popped out of a mold in four minutes while a composite shell takes a whole day. And it takes a whole 'nother day to paint the damned thing," says Simpson. And then there's the price. "The material cost for a composite is probably twenty times the cost to make a plastic piece."

Simpson's approach was classic applied engineering: high-level trial and error. "I used all kinds of composites and did the same thing with liners," he recalls. "And I ended up with a composite of three different foams, each doing a different thing."

These advanced materials also made Simpson's helmet lighter than existing models, which can help a middle linebacker's head be more like a woodpecker's: less massive and thus less prone to concussion. "I hear [scientists say], 'Linear accelerations don't really have a lot to do with it. It's these *rotational* head injuries [that cause concussions].' The only way I think you could mitigate that would be to make a lighter-weight part," says Simpson. "We set out to do that, and we make a part that weighs less than two pounds."

And before long, Simpson had created a promising prototype.

"When I finished with this thing after four hundred prototypes and a couple thousand drops, I had what I thought would be a great product," says Simpson. "And that's where we are now."

In his 1959 lecture "The Two Cultures," C. P. Snow argued that the "intellectual life of the whole of Western society" was separated into two camps—the sciences and the humanities. And the cultural chasm between them is so vast that it's preventing many of the world's problems from being solved. If he examined the issue of helmets and concussions, he might have recognized a similar rift between the scientists and the engineers.

Like Lorna Gibson and Bill Simpson, scientists and engineers approach the world very differently. "Scientists discover how nature works," says Chuck Vest, president emeritus of MIT and president of the National Academy of Engineering. "And engineers create new things."

While most people tend to lump together everyone who wears a lab coat, engineers and scientists have very different goals and approaches, and they often don't see eye to eye. "Look at *The Big Bang Theory*," says Norman Fortenberry, executive director of the American Society for Engineering Education. "They constantly make fun of how [Howard] is only an engineer without a Ph.D. They pick on him."

These differing worldviews may ultimately be working against each other in the quest to create a football helmet that better protects players from concussions and head injuries.

"Scientists are seeking to understand why this is happening," says Fortenberry. "An engineer is focused on 'How can I prevent it?'"

In the field of head injuries, scientists have a lot to try to understand as they parse the puzzle of concussions and the related long-term degenerative brain disease known as chronic traumatic encephalopathy, or CTE.

Just how does a concussive impact impair the function of the brain?

"You've got this metabolic crisis going on within the cell," posits Boston University's Robert Cantu. As potassium ions flood out of the nerve cell, he explains, they're replaced by calcium ions, which prevent the cell from passing on information.

Is there a genetic component to concussions and CTE?

"No one knows yet," says Robert Stern, a professor of neurology at Boston University, "but studies are focused on a variant of a common lipid transport gene called ApoE-ε4. This gene does good things, making sure fat goes to the right place, but if you have the wrong form, it does something crazy in the brain." Stern adds that the ApoE-ε4 is a susceptibility gene, as opposed to a deterministic gene. "If you have the wrong form, it increases your risk of having the disease, but it does not mean you will get it," he explains. "There is not going to be a CTE gene because it is such a multifaceted disease."

And how do the more frequent sub-concussive hits contribute to CTE?

"The big hits have a bigger effect, but how many little ones do you need to be worse than a single big one? We don't know," says Cantu. "My guess is that in ten years we will know: Does a hit of 100 G equal ten hits of 10 G, or five hits at 20 G, or is it ten times worse than a hit at 50 G?"

And while the details of these questions can be kind of dry, and some of them will lead only to knowledge purely for knowledge's sake, let's remember that those with the most at stake—the players battling this disease—see this kind of scientific research as a source of hope even in their darkest days.

In 2010, for example, former Chicago Bear Dave Duerson, who was suffering from the symptoms of brain trauma, committed suicide; he was careful to shoot himself in the chest so that researchers could study his brain and examine it for signs of CTE. The brains of thirty-five other players have been donated to Boston University's Center for the Study of Traumatic Encephalopathy; alarmingly, CTE was found in thirty-four of them.*

This kind of basic research, however, takes time. Engineers, on the other hand, aren't particularly patient when it comes to running decades-long longitudinal studies. They argue that athletes are suffering both long- and short-term injuries right now, so even an incremental improvement in helmet design could save brains—and lives.

An extreme example of an engineering-based approach can be found in *Apollo 13,* the capsule's carbon dioxide scrubbing equipment was failing and, with the clock ticking, a group of engineers had only hours to improvise a solution that included fitting a round peg into a square hole—or, actually, fitting a round air hose into a set of square filters. The NASA engineers cobbled together an improvised fix using a few binders, plastic bags that would have been used to collect moon rocks, a hose, a pair of socks, a bungee cord, and duct tape.

Engineers tend to be very goal-oriented. The clearer the target, the more easily they can focus on hitting it. When it comes to concussions, Simpson is frustrated by the lack of solid answers to even his most basic questions. "No one knows what a concussion is. That is not a joke," he says. "I've asked these neuroscientists, 'What is the threshold for a concussion?'"

Scientists, for their part, are equally frustrated by engineers asking questions to which there are no easy answers. In practice, the range of impacts that might result in a concussion stretches between 30 G on the low end up to 200 G on the high end, depending in large

---

* CTE is marked by the presence of tau protein in nerve cells of the brain, which can be determined only during an autopsy. In 2013, a small-scale study at UCLA used PET scans in an attempt to identify the presence of CTE in five living former NFL players. "This is the Holy Grail if it works," said Cantu. "This is what we've been waiting for, but it looks like it's probably preliminary to say they've got it."

part on the individual, as well as sheer luck. "You have the biomechanists, and they are hung up on linear and rotational force," says Cantu. "They are big on the numbers. That is their life. That is their training. The problem is that's only half the problem. We are dealing with human beings [so we need to ask questions like]: How many concussions have you had? How severe? How close together? How easily are you able to be concussed?"

Engineers point out that they often come up with solutions before the science is fully in place. A prime example is the steamboat, which was carrying passengers up and down America's rivers long before scientists fully understood the theory behind steam-powered propulsion. "We didn't have thermodynamics [back then]," says Fortenberry.

Are engineers and scientists doomed to perpetual conflict, the Odd Couple of the laboratory? Or can engineers and scientists find ways to work together without driving each other crazy? Vest says yes, they can work as a team. "In the more applied areas, like use-inspired basic research, this [chasm] is starting to get blurred between what scientists do and what engineers do," Vest explains. He points to engineers working side by side with biologists in Phil Sharp's cancer research laboratory at MIT. "They are working intellectually together, because the engineers have an understanding of how systems work differently from biologists, and this is turning out to be really productive."

Studying woodpeckers may lend insight into the subject of head injuries, but not much of it can be put into practice in preventing concussions in humans. But a larger animal known for its hard head might help us to do just that. Rams, or male bighorn sheep, are big animals with relatively big brains. During their mating season, they engage in a head-butting ritual, during which they collide head-to-head at speeds between 20 mph and 40 mph. They hit each other with tremendous force but do little damage.

How do they survive?

Their horns.

A ram's horn is made of a porous bone that is covered in keratin, a protein found in our hair and nails and elsewhere in the animal kingdom in claws, tortoise shells, and porcupine quills. Keratin is stretchy and elastic, which allows the horns to give a bit under impact. The keratin layer can have a thickness ranging between a few millimeters and a couple of centimeters. And it has the much-needed shock-absorbing properties to reduce the impact.

Andrew Farke, a paleontologist who has studied rams and horned dinosaurs, explains: "This sheath of keratin is very, very thick in animals that head-butt. And the great thing about keratin is it deforms easily." This deformation helps to distribute the force of the impact by providing a greater surface area. Ram horns also bend backward during a collision, which also helps reduce the force.

There is a concept in physics called the impulse, which is equal to the force of the impact multiplied by the time of the impact. The impulse is constant, so if you increase the amount of time, the force experienced will decrease. It's easy to demonstrate this concept by dropping an egg. When an egg falls on the concrete, the time of impact is very short, which causes the shell to crack. Dropping an egg onto a cushion increases the duration of the impact as the egg sinks into the cushion and springs back up. And the shell remains intact.

When they're butting heads, rams are going from more than 20 mph to zero at impact. The bending of the horn increases the duration of the impact, which lowers the force, and that allows the brain to slow down over a longer period of time. The horns are acting a bit like the crumple zones in a car. But in a car, the crumple zone is designed to sacrifice itself during the crash. Ram horns are not. Like football helmets, horns need to be ready to absorb the force of repeated impacts.

A helmet adapting the design and materials from a ram's horn might provide an answer to pro football's head injury issue. But

Farke isn't in contact with the NFL. "There is a disconnect. We could probably talk to each other more," he laments. "I am a paleontologist. I mainly deal with dead stuff."

Bill Simpson has a prototype helmet that's lightweight and tests well at absorbing impact. He also has a track record of building products that protect race car drivers from concussions at speeds of up to 250 mph. And yet, when he looks back on his attempt to reinvent the football helmet, his main feeling is frustration.

"It's a cesspool of politics, this whole thing," he says. "I'm trying to do the best I can do, but I'm being ignored."

Not a day goes by that the NFL isn't contacted by a dozen workbench inventors with homemade helmets that will prevent concussions. But few, if any of them have Simpson's credentials and experience. "I've written letters to Roger Goodell, and they get ignored. They don't want to know about my helmet. It's just one bad thing after another," Simpson continues. "I don't know what drives this whole thing. I just don't know." And so Simpson has given up trying to get his helmet used by NFL players, shifting his attention instead to models for the youth market so twelve-year-old kids can play the game more safely.

Let's not confuse disturbing news with bad news. There's much to be pessimistic about in pro football's head injury situation. The helmets currently in use don't seem to protect players particularly well from concussions. A promising alternative isn't getting much attention. Scientists and engineers often seem to be working at cross-purposes. We don't really know enough about the causes of head injuries or the ways to prevent them. And doing enough research to build up that body of knowledge could take years, if not decades, and will cost plenty of money.

But there's a silver lining. Things could be worse. If the helmets we have do a fabulous job of protecting the head, and we have an almost complete understanding of head injuries and what causes them. And despite all that, players are still suffering concussions and getting CTE. That would mean there's very little chance of making the game safer through science and engineering. But, as it is, that *isn't* the case. As it is, there *is* a chance things can get better.

Football faces a lot of problems. And also has a real opportunity to make the game safer through science or engineering. Viewed through that prism, our current lack of knowledge could prove to be football's salvation. In ignorance lies hope. And that hope will grow exponentially if scientists and engineers can find a way to work together as they search for the answers.

SHOULD THE NFL BAN HELMETS?

In 2009, the Insurance Institute for Highway Safety decided to do an experiment.

The IIHS is a nonprofit agency funded by the insurance industry that works to increase automobile safety. One of their highest-profile activities is crash-testing late-model cars into solid objects. To commemorate the organization's fiftieth anniversary, they decided to do something different. They would test a 2009 Chevy Malibu against its fifty-year-old equivalent, a 1959 Chevy Bel Air. In this test, the vehicles would crash into each other. While many automobile purists think of older cars as being built like proverbial tanks, the IIHS test told a different story.

Both cars were severely damaged. But the new Malibu's crumple zones absorbed much of the impact, and the air bags, head restraints, and seat belts controlled the movement of the driver. There was a mess, but it was a controlled mess. The Bel Air was a different story. The sheet metal morphed into something that looked like modern art, but most important, there was significant intrusion into the passenger compartment. The Bel Air driver would likely have suffered serious chest, head, and neck injuries, while the Malibu driver would have escaped with nothing more than a foot injury. Or to put it another way, the Malibu driver would have hobbled away from the crash while the Bel Air driver would have left in a medevac helicopter. Or maybe a body bag.

"It was night and day, the difference in occupant protection," said IIHS president Adrian Lund. "What this test shows is that automakers don't build cars like they used to. They build them better."

Inspired by this test, Adam Bartsch, a biomedical researcher at the Cleveland Clinic, decided to apply its methodology to football helmets. He'd take a seventy-year-old leather helmet and a brand-new model and do roughly the same thing. Mimicking the IIHS Malibu versus Bel Air smackdown, Bartsch crashed one helmet against the other, in a way that's typical of the kinds of impacts that take place in a real football game.

What he discovered was more than a little disturbing.

"Why Leather Football Helmets Could Provide a Better Defense Against Concussion" blared the headline on Time.com. Even more disturbing was the more nuanced explanation found in the text of Bartsch's paper. It's not that the leather helmet was surprisingly *good* at protecting players from concussion. It's that the modern helmet was surprisingly *bad*.

Which prompts this question: Should the NFL ban helmets?

The question is not as ridiculous as it might seem at first glance. Other contact sports are played without helmets: rugby, Australian rules football, soccer. In the early days of American football, the game was played without helmets. There haven't been studies that directly compare American football to non-helmet contact sports when it comes to head injuries. However, a 2001 study by researchers at the University of North Carolina at Chapel Hill reported a concussion rate for high school rugby players of 3.8 per thousand athlete exposures. A report by the Australian Football League Medical Officers' Association cites a concussion rate of between 5 and 6 per thousand player hours at all levels of the sport. Researchers from the University of Georgia found that college football players had a concussion rate of 11.1 per 1,000 athlete exposures over a sixteen-year study. While this is not a clean comparison, the fact that the incidence of concussions is two to three times as high for American football provides another piece of evidence that helmets may have limited benefits in reducing this kind of brain injury.

A pro football Hall of Famer makes a similar argument about restricting protective gear. Mike Ditka, a legendary player and a Super Bowl–winning coach, has argued in favor of banning the face mask. "I said a long time ago if you want to change the game, take the mask off the helmet," he said to *USA Today*. "It will change the game a lot. If you want to change the game and get it back to where people aren't striking with the head and using the head as a weapon, take the mask off the helmet. A lot of pretty boys aren't going to stick their faces in there."

Ditka's sentiments echoed those of the late Joe Paterno, who also argued for the elimination of the face mask in 2008, long before controversy ended his tenure at Penn State. "I have been saying [it] for fifteen years," said Paterno, who had played during the 1940s' leather helmet era and saw the radical changes in the game as equipment evolved. And while he surely knew that banning the face mask was a pipe dream, he still talked it up. "Then, you would get back to shoulder blocking and shoulder tackling, and you wouldn't have all those heroes out there. Guys would have to worry about broken noses, knocked-out teeth, which we would like to prevent, but you don't get anything for nothing. We used to have one single bar. Now we have a weapon."

One of the first things you do when you write a book like this is start asking questions. Lots of them. From reasonable questions like "How did football players get so big?" to others that are totally bonkers, like "What would happen if you played football underwater?" The helmet question was one of the very first questions we asked.

Should the NFL ban them?

Is it crazy? A little. But there's value in craziness.

Asking provocative questions, that's what scientists do. Before they ever set foot in a lab, scientists ask questions. Big ones. Little ones. Smart ones. Stupid ones. And most of all, questions that lead to yet another question. This Socratic give-and-take is at the heart of the scientific method. And it's at the heart of this book.

That's the reason why scientists in a whole variety of fields were willing to talk to us, even though their research didn't concern football, and in many cases they hardly knew anything about the game. What they did understand was the process of asking and answering questions. On some level, *Newton's Football* is about finding the common ground between Isaac Newton and Vince Lombardi, between Bill Walsh and Erwin Schrödinger. But on another, it's simply about asking crazy questions that lead to unexpected answers. It's what scientists do. The best minds in football do it, too.

A football stadium is a laboratory with body paint and foam fingers and seven-dollar hot dogs. We've seen how even tiny tremors can shake the game to its roots. Like Otto Graham's face mask. Like Greg Cook's shoulder injury. Like Bobby Hosea's worry about his young son getting hurt. And maybe that's why hundreds of thousands of fans gather at stadiums every Sunday, as millions more crowd around their televisions, all for the chance to see something new and unexpected. Like Isaac Newton waiting breathlessly for a Falling Apple or a particle physicist chasing the next Higgs boson, football fans are perched on the edges of their seats, anticipating that next installment of innovation in action.

So what about banning helmets? The logic seems simple enough. Football has a concussion problem. Helmets don't seem to protect players particularly well against concussions. In other sports, where the players don't wear helmets, there doesn't seem to be much of a concussion problem. And football didn't have a concussion problem back in the day before helmets were mandatory.

Ergo, ban the helmet. End concussions.

Not so fast. Robert Stern, the eminent Boston University researcher, was among those who wrote a vehement response to the *Journal of Neurosurgery* about Bartsch's article. Stern and his colleagues were concerned, with good reason, that the mainstream media would report only half the story, writing sexy headlines about a return to leather helmets and adorning the article with production stills of George Clooney from *Leatherheads*.

Stern's response to the larger question of banning helmets is equally pointed. "Remember that the reason why helmets were created in the first place was to prevent skull fractures," says Stern. "So if you get rid of them, you are going to have people die. That's why that's not the answer."

The most profound thing we learned in writing this book is that football is a complex, dynamic system. We've seen how the solution to one problem gives rise to another. Solve the problem of fractured jaws and broken noses, and up pops the concussion problem. Solve the concussion problem, and skull fractures come back. It's a game of neuroscience Whac-A-Mole.

But it doesn't end there. Banning helmets would also change the way that players play the game. Some of those changes might reduce the chances of players suffering concussions and skull fractures. Others would surely increase them.

"In some ways it does make sense," Stern suggests. "Would it be good to change the helmet or take them away so [players] feel pain and it wouldn't be easy to bump heads over and over again? Yeah, probably. To get rid of that false sense of protection would be sensational."

Professional football has never been more popular than it is right now. At the same time, the growing concern about head injuries has blossomed into a larger uneasiness about the future of the game it-self. Pundits like Malcolm Gladwell have compared football to dog fighting, while more moderate voices have warned that the sport could become marginalized in the manner of professional boxing.

But as we've seen throughout the pages of this book, it's not the first time pro football has been in a place like this. The elements that make the game appealing and the elements that make it dangerous have always been one and the same. "Dancing is a contact sport," said Vince Lombardi. "Football is a *collision* sport."

And so the game sits at a crossroads.

Or perhaps something else.

In 1962, Thomas Kuhn wrote *The Structure of Scientific Revolutions*. It's not a science book but a book *about* science, about the way science works. Until Kuhn wrote *The Structure of Scientific Revolutions,* it was accepted that scientific knowledge progressed in an orderly fashion: each new theory and each new experiment adding incrementally to an existing body of knowledge.

Kuhn argued that, in reality, change actually took place in sudden, even violent, shifts. He called them paradigm shifts, and you can see them at work throughout the history of science, from the Copernican revolution in astronomy to Einstein's theory of relativity. We take them for granted now, but at the time these battles were hard fought and hard won. Kuhn also posited that this principle works not just in the lab but in the world at large.

Including the world of football. The history of pro football has been forged through a series of these Kuhnian paradigm shifts, battles between those who would tame the game and those who would find a way around those civilizing influences.

Kuhn also argued that every paradigm shift is preceded by a crisis. A crisis much like the one that the NFL is experiencing now. If he were here today, Kuhn might contend that football's future is only a paradigm shift away.

Maybe it'll come in the form of a new helmet.

Or a new way of tackling.

A new offense.

Or a new player.

A new coach.

Or a new idea.

Will that idea be banning the helmet?

Probably not.

But it might just come in the form of an answer to some equally crazy question.

4-3 DEFENSE: A defensive arrangement featuring four lineman and three linebackers. Other variants include the 3-4 defense, the 5-2 defense, and so on.

AUDIBLE: A play call initiated by the quarterback at the line of scrimmage by vocalizing the new play, usually using some code that his teammates understand. Some examples are: "Mike 52!" where Mike means middle linebacker and his jersey number is 52. "London Bobby!" where London means players are going to run to the weak side on the left, because the tight end is on the right.

BLITZ: A sudden attack from several defensive players who abandon pass coverage and rush the quarterback to disrupt the offensive play.

BOMB: A long pass.

BUMP-AND-RUN: A defensive strategy in which a cornerback lines up with a wide receiver and impedes his path or throws off his timing to disrupt him from running his route. This is only allowed within the first 5 yards from the line of scrimmage.

COMBINE: The NFL's annual scouting event, generally held in Indianapolis, that showcases potential players by testing their physical traits. Players are evaluated for the 40-yard dash, the bench press (number of repetitions of a 225-pound weight), jumping (vertical and broad), shuttles (20 and 60 yards), and three-cone drills. Certain positions have additional tests, including intelligence tests.

CORNERBACK: Defensive player who covers pass zones on the outer edges of the field.

COVERAGE: A defensive strategy used to prevent a pass completion.

DIME: A defensive arrangement with six defensive backs, usually used to respond to passing plays.

GO ROUTE: A route in which the receiver runs straight toward the end zone, with the intention of outrunning the defensive players to catch a pass for a potential touchdown.

HANG TIME: The duration a football stays in the air after it has been kicked.

LINE OF SCRIMMAGE: An invisible boundary that neither the defense nor the offense can cross until the play begins.

NICKEL: A defensive arrangement in which five defensive backs are used, usually to respond to passing plays.

NO-HUDDLE OFFENSE: Also called the hurry-up offense. An offensive strategy of using as little time as possible between plays. Shortening the time prevents the defense from using appropriate players or strategies for the next play.

POCKET: The area created by the blockers to protect the quarterback.

PRESSURE: Disruptive intimidation brought about by imposing fast movement or lots of bodies toward the ball.

REDSHIRT: A person who sits out an entire football season to retain eligibility later on in their college career.

SECONDARY: Cornerbacks and safeties—pass defenders who play behind the linebackers and defensive linemen. This term also refers to the area of the field they occupy.

SNAP: The quick transfer of the ball from the center to the quarterback that sets the play in motion. It designates the start of the play.

STRONG SIDE: The side of the offensive line where there are more players lined up. It's usually, but not always, the side where the tight end is present. (See *weak side*.)

WEAK SIDE: The side of an offensive line where there are fewer players lined up. It's usually, but not always, the side where the tight end is not present. (See *strong side*.)

WONDERLIC TEST: A multiple-choice intelligence test used in the NFL in assessing draft picks.

ZONE BLITZ: A technique of applying defensive pressure by combining the blitz with zone coverage.

ACCELERATION: The rate of increase or decrease in speed. It is measured in meters per seconds squared (m/s^2) and is designated as a.

ANALYTICS: The harvesting and examination of data to cull and determine meaningful patterns or to draw new conclusions.

APOLIPOPROTEIN E (APOE): A protein that transports lipids (fats and cholesterol) through the bloodstream to neurons after an injury. The human body has three variations designated with the Greek symbol epsilon (ε); they are ApoE-ε2, ApoE-ε3, and ApoE-ε4. ApoE-ε4 has been specifically linked to concussions (and mild traumatic brain injury) and is associated with increased risk of Alzheimer's disease. About 13 percent of the general population has this gene. It is a susceptibility gene and not a deterministic one.

BINARY SYSTEM: A system that consists of two numbers or modes (1 or 0, on or off, left or right, up or down). Microprocessors are made up of switches with two modes, 1 and 0. The binary system is the language of the computer and is used to calculate and store data. Binary is also called a base-2 system, meaning that there are only two options (1 and 0) for each numbers place. Unlike the more common decimal system, which positions digits in the ones place, the tens place, the hundreds place, and so on, base-2 has units in the ones place, the twos place, the fours place, and so on. So the binary number 0010 would represent 2 in a base-10 system, and 0100 would represent 4.

CHAOS THEORY: A field of mathematics that has been applied to physics, biology, and economics. It studies the behavior of systems that are very sensitive to their starting points or initial conditions. (See *double pendulum*.)

CHRONIC TRAUMATIC ENCEPHALOPATHY (CTE): A neurological degenerative disease of the brain caused by trauma. The symptoms include dementia, memory loss, and depression. In the early twentieth century, this condition was called "punch-drunk" and was found in a number of boxers who ultimately were discovered to have suffered from dementia or Alzheimer's disease. No cure for CTE is currently known, and at present it can only be identified postmortem.

COMPLEXITY: A term referring to a system with many parts and many linkages between these parts. The science of complexity is the study of the behavior and relationships that arise from these interacting parts. Complex systems include telephone networks, ecosystems, brains, cities, and economies.

CONCUSSION: A type of mild traumatic brain injury that occurs through a jolt of the head caused by a direct impact or a blow to the body and can cause the brain to undergo impact with the skull. This, in turn, strains the brain cells and impairs their functioning. Symptoms include headaches or loss of balance, memory, concentration, coordination, and judgment. The CDC estimates that 1.6 million to

3.8 million people suffer sports-related mild traumatic brain injuries in the United States every year.

DOMINANT STRATEGY: In game theory, this is the best strategy a player in a game or decision-making process can choose. It is the strategy a player would undertake even if they knew what their competitor was planning.

DOUBLE PENDULUM: A pendulum with another free-moving pendulum on its end. When you start the swing, the pendulum will have an erratic or chaotic motion. Starting the pendulum at the same place will not generate the same path of motion, which shows the strong sensitivity to initial conditions—a demonstration of chaos.

$F = ma$: This formula indicates that the acceleration of a body, a, is directly linked to the force that is applied, F, but inversely linked to the mass of the body, m. Also known as Newton's Second Law of Motion, it means that the more mass a body has, the more force is going to be needed to accelerate it.

FAIRING: A structure with the primary purpose of reducing drag.

FRAGILITY: A vulnerability or "Achilles' heel" to a previously overlooked condition.

GAME THEORY: A branch of mathematics that analyzes the best strategy for a player who is interacting in competitive situations. The decisions take into account rewards (payoffs) and the amount of information each player has.

GENES: The discrete units that make up our chromosomes and give instructions to make proteins. They determine physical traits and heredity and are made up of DNA. Mutations occur when a DNA sequence in a gene changes.

MOMENTUM: The velocity of an object multiplied by its mass, designated as p in the following formula: $p = mv$. It is mass in motion and was described by Newton as the "quantity of the motion." Momentum is always conserved, and is transferred from one body to another.

NASH EQUILIBRIUM: A strategy from game theory in which no player has an incentive to change the chosen strategy after considering the opponent's choices. Players receive no additional rewards (payoffs) from changing their strategies. It is the optimum strategy of a game.

NATURAL SELECTION: As noted by Darwin, this is a process that can lead to the evolution of new species. In a population, there is a large variation of traits. When conditions such as changes in the climate or environment, competition for food, disease, or other factors occur, a percentage of the population with traits that are less suited for these new conditions will perish. The better-suited part of the population survives and passes on these traits to their offspring. Over time, this process can give rise to species that are remarkably different from the original population.

OBSERVER EFFECT: A term that refers to the fact that the act of observing a process can sometimes alter the process that is being observed. Measuring the pressure from a tire requires that you let a bit out, which changes the tire pressure slightly. Opening the oven door to look at how your pot roast is doing causes the temperature to decrease a bit. In many cases, interference by observing cannot be avoided, so it must be taken into account when describing an event.

PRESSURE: A force applied on an area.

PROLATE SPHEROID: A sphere that has been stretched more in one direction than the other. Common prolate spheroids are watermelons, rugby balls, and the American football.

PUNCH-DRUNK: Known among medical professionals as *dementia pugilistica*, this is the same condition as chronic traumatic encephalopathy, which was first associated with boxers. Dr. Harrison Martland, a pathologist and the chief medical examiner for Essex County, New Jersey, coined the term "punch-drunk" in the 1920s.

QUANTUM THEORY: A field of physics that describes the workings and behavior of atoms and electrons. Their behavior departs significantly from large objects like a cannonball. First off, electrons can behave like a wave (like at a beach) and like a particle (like a marble). When an electron travels through the air, it acts like a wave and is more nebulous to describe. When an electron crashes into a wall, it acts like a marble. These contradictory behaviors are the cornerstone of this scientific field, which describes the behavior of particles and their properties. Leaders in this field include Erwin Schrödinger and Werner Heisenberg. While Einstein initially supported the notion of quantum mechanics, he thought this theory would be replaced by something better down the road and often took a contrarian position to some of its claims.

RANDOMNESS: A term used to describe a lack of predictability of an event.

ROBUSTNESS: A term that describes the ability of a system or object to resist change despite experiencing new or unusual conditions. A robust system or object continues to function the same way despite changing conditions or environments.

SCHRÖDINGER'S CAT: A thought experiment proposed by Erwin Schrödinger to show the weirdness of quantum mechanics. If a cat is placed in a box with radioactive material that has a 50/50 chance of emitting a particle, there is a 50/50 chance the cat will be dead or alive. But while the box is closed, an observer cannot know which state the cat is in. Both states, which are vastly different, occur at the same time. The dead-alive cat shows how quantum mechanics can have two very different states superimposed on each other. This is called superposition and is a mind-boggling difference from common sense. We commonly understand objects to be in either one state or the other but not both. In quantum mechanics, however, you can have both.

THERMODYNAMICS: A field of physics concerned with heat, its conversion to other forms of energy, and its ability to do work (that is, move an object to a new location). It concerns itself with changes in pressure, volume, and mass. One important law is that systems slowly move to disorder, that is, increase in entropy. Thermodynamics was used to understand how steam expands when heated, which was later used to explain the workings of a steamboat. Thermodynamic principles can be applied to understand the motion of objects from molecules to galaxies.

UNCERTAINTY PRINCIPLE: A statement about the limits on our ability to describe the behavior of subatomic particles. When speaking about objects that are atomic-sized or less, the precision of our information becomes less and less accurate. We cannot know things precisely but can know things in terms of probabilities. On a large scale, one can know the position and the velocity (actually momentum) of an object, such as a bowling ball, very well. But for an electron, one can know either the position or the momentum with certainty, but not both. As a person gains more precision with one, he or she loses it with the other. A bowling ball is described by Newtonian (classical) physics. An electron is described by quantum physics, which denies the possibility of such precision. That is the funky world of quantum physics!

VISCOSITY: A measure of the resistance to movement a fluid (that is, a liquid or gas) has when a force is applied.

WABAC MACHINE: The fictional time travel device used by the cartoon characters Mr. Peabody and Sherman on *The Rocky and Bullwinkle Show.*

ACKNOWLEDGMENTS

If you're lucky as a writer, you find an editor who'll buy your book. A far rarer thing is an editor who *gets* your book. Someone who not only likes what you wrote, but who truly understands it.

That's Mark Tavani. He understood what *Newton's Football* should—and could—be even before we did. He was not only our biggest fan, but our smartest critic, too. He's also a lot funnier than he thinks he is.

Thanks also to Betsy Wilson, who's been an indispensible team player on Team Newton. Thanks to copy editor Martin Schneider for catching mistakes big and small, and to Ted Allen and the rest of the Random House production team for helping turn these words into a book. We're big fans of Wes Youssi, who did the brilliant cover design for *Newton's Football,* and Dana Leigh Blanchette, who designed the lovely pages.

A book's not worth much if you keep it a secret. Cindy Murray, Greg Kubie, and Quinne Rogers are the heart of Random House's crack publicity and marketing team. They're consummate pros and they've done a bang-up job of telling the world about *Newton's Football*.

Danny Tuccitto, our technical editor, was a remarkable late addition to Team Newton. From enlightening us on the finer points of the zone blitz to nudging us when we got a little too familiar with a Nobel Prize winner, Danny made this book better in more ways than we can count. Look for Danny's very smart writing at *Football Outsiders* and elsewhere. Thanks, too, to Aaron Schatz, who introduced us to Danny.

What's better than a caper flick where they get the old gang together? Allen's friend and editor par excellence, Brian Parks, joined Team Newton for a late lap through the pages.

We'd like to extend our gratitude to Alan Schwarz, who introduced us to Bobby Hosea and otherwise shared his expertise on head injuries and football helmets, a subject about which he knows more than just about anyone. His reporting at *The New York Times* has moved on from sports to more serious subjects, and the world is a better place for it.

Thanks to author Stefan Fatsis, who shared research material from an unpublished chapter of his splendid book on placekicking *A Few Seconds of Panic: A Sportswriter Plays in the NFL*. It should be next on your reading list.

Thanks to Wayne Coffey of the *Daily News,* who helped us in our search for Ben Agajanian; Dan Daly, who hooked us up with Fred Bednarski; and John Breech, who got his dad to talk to us. Twice.

And a shout-out to Moira McCarthy Stanford, Mike Rapaport, Fredric Alan Maxwell, Jerry Beilinson, and Al Mercuro who all read drafts of the manuscript and lent us some needed perspective.

And thanks to all those—some quoted in the text and some not—who shared their knowledge and insight with us. That very long list includes: Jerry Rice, John Nash, A. Douglas Stone, Jani Pallis, Marvin Chun, Verge Ausberry, Matthew Jackson, Roberto Serrano,

Julian Walshaw-Vaughan, Timothy Cowan, Toan Pham, Ryan Morse, William Conine, John Matson, Bobby Hosea, Richard Steckel, Lorna Gibson, Robert Malina, David Kaiser, Steven Tommasini, Tuomo Rankinen, Dick Zare, Virgil Carter, Charles Greer, Mark Gerstein, Harper Reed, Ryan Tierney, David Maraniss, Robert Stern, Kurt Bryant, Steven Humphries, Sam Wyche, Bill Simpson, Tom Bass, Daniel Gopher, Jocelyn Faubert, Charlie Ward, Chuck Yesalis, Jay Hoffman, Robert Cantu, Chuck Vest, Norman Fortenberry, Paraag Marathe, Brian Burke, Jonathan Weiner, David Katz, Al Toon, Fred Bednarski, Don Shula, Keith Chen, Steve Wolfram, Sian Beilock, Andrew Farke, Mark Saba, and Andy Shimp.

AINISSA RAMIREZ'S ACKNOWLEDGMENTS

First off, I want to thank my family: my mother, Angela; my brothers, Marc and Davyd; my sister-in-law, Cassandra; and my niece and nephew, Lena and Alex. My brother Marc deserves a special shout-out, since he was invaluable to this project by translating football terms for me. He made this book better. My friends have been the best, as they always seem to support my crazy ideas. They are Sylvia Stell, Sarah Marxer, Robin Shamburg, Wendy Sealey, Margot Abels, Kathy Yep, Andrea Queeley, Rene Miranda, Erin Lavik, Jonny Skye, Sandra Brown, Priya Natarajan, Ajuan Mance, Tammy Rogers, Sandy Yulke, Walter Gray III, Demetrius Eudell, and Eric Hayes.

A big hug goes to my science teachers, who instilled and nurtured my love for science. They include Kathleen Donahue (grammar school science teacher), Jean-Marie Howard (high school science teacher), Edelgard Morse (college chemistry instructor), Prof. Clayton Bates (my grad school mentor), and Dr. James Mitchell (my Bell Labs mentor). I've also benefited from the tutelage of the late Stig Hagström, Robert Sinclair, William Nix, and Mike Kelly.

A shout-out goes to my science communicator running buddies

who keep me on pace: Emilie Lordtich, Leslie Kenna, Katherine Vor-volakos, and AFI. My students, who inspire me to be a better teacher and human being, deserve mention; some of them include Katie McKinstry, Guy Marcus, Jeremy Poindexter, Tania Henry, and Xu Huang.

I have benefited from mentors, sages, and supportive folks and organizations that include Samuel Allen, Carol Lynn Alpert, Janet Keller, Alison Chaiken, Lisa Marcus, Usha Kanithi, Milton Chen, Anne Fausto-Sterling, Robin Rose, Mary Lou Bednarski, Deborah Proctor, Alan Schwartzman, Ron Knox, David Johnson Jr., Victorio Sweat, Gary Gates, Nancy Santore, Mike Hughes, Ray Matthews Jr., Shirley Malcom, the folks at TED, Charley Ellis, Neil Tyson, Carl Zimmer, Gina Barnett, Robin Hogen, Karen Peart, Helen Dodson, Steve Geringer, Ms. Terry, the late Mrs. A. Green, Marie Fabrizio, Paul Fleury, and Peter Salovey.

My literary agent, Laura Wood of FinePrint Literary Management, has been unceasingly supportive in getting me from zero to hero in a short amount of time. Thank you, Laura! I have also benefited from working with the Jodi Solomon Speakers Bureau in helping me spread my message that science is fun.

Also, my special thanks goes to Allen St. John for lending his beautiful mind to this book project inspired by Sir Isaac Newton. Thanks, Allen!

To be honest, there are far too many people who helped me along the way, so let me end with a broad but heartfelt "Thank you!"

ALLEN ST. JOHN'S ACKNOWLEDGMENTS

"Pop had Genco. Look what I got." That was what Sonny Corleone said in frustration as he prepared for war with the Tattaglia family in *The Godfather*. He was wondering why he couldn't have a wartime consigliere. Me? I don't wonder. I've got Jason Allen Ashlock of Moveable Type Management. Most of the time, my agent is all smiles, but when it's time to go to the mattresses, his advice is trustworthy. And so is he. *Molte grazie, Consigliere.*

Thanks also to Dan Gerstein of Gotham Ghostwriters, who, through the roundabout magic of Facebook, introduced me to Jason.

Big thanks, as always, to all my editor friends who give me something to write about between book projects, especially my new buddies at *Road & Track,* Larry Webster, Sam Smith, John Krewson, and Josh Condon. As well as Mike Fazoli who might actually read this book. A special shout-out to Jim Kaminsky with whom I've shared many an adventure and many a late night. And my writer friends who helped make this book a little smarter: Allen Barra, Charlie Pierce, Bill Barnwell, Jonah Keri, and Peter Richmond.

A big "Hey, how you doin'" to Wayne Henderson, who again lent me his spare room during crunch time, and inspired me with his peerless work ethic and dedication to craftsmanship. A shout-out to the denizens of the guitar shop on Tucker Road, Elizabeth Henderson, Herb Key, Don Wilson, and Harrol Blevins, as well as the memory of Dave Neal and Ralph Maxwell. And to T. J. Thompson, who's always encouraged me to work hard and think harder.

And then there are the reasons I write and, for that matter, the reasons I get out of bed in the morning: my wife, Sally; my son, Ethan; and my daughter, Emma. Writing a book is a team effort, and this time they really took one for Team Newton. Dad missed a few travel softball games, a few regattas, and more than a few family dinners, but every word, every letter, every space is for you. I've got a lot of words, but none to express how much I love you.

Same goes for my favorite nieces, Nicole and Casey, my new favorite nephew, Tim, and John and Abby, who one day soon will be old enough to read this.

And a moment of silence for Marvin the Mouse, ounce for ounce the best pet ever. Thanks, too, to Team Newton's official mascot, Caymen's Comet Full of Love aka Tessie the golden retriever. We all need a best friend.

Last but not least, a huge thank-you to my co-author, Ainissa

Ramirez. From the day we formed Team Newton, she's been this project's biggest cheerleader, and I hope these pages give you a sense of her big heart. And her big brain, too. Business types throw around the words like "brainstorming" a lot, but that's exactly what was at the heart of this book: Ainissa telling a science story, me telling a football one, and searching for the connections together. It's a rare and beautiful thing. Thanks. 16,632 times.

NOTES

INTRODUCTION

xiii **"ant-like entities"** Stephen Wolfram, phone interview by the authors, January 13, 2013.

xiv **That morning last January, we had called** Sam Wyche, phone interview by the authors, December 24, 2012, and April 9, 2013.

CHAPTER 1: THE DIVINELY RANDOM BOUNCE OF THE PROLATE SPHEROID

3 **The Philadelphia Eagles Pro Bowl** John Branch, "Eagles Stun Giants on Game's Final Play," *New York Times*, December 19, 2010.

4 **A herd of Giants defenders** Andrew Perloff, "Eagles' Rally Causes More Giant Heartache, More Lessons Learned," *SI.com*, December 19, 2010. Available at http://sportsillustrated.cnn.com/2010/writers/andrew_perloff/12/19/five.things.eagles.giants/index.html, accessed July 2, 2013.

5 **It's called the Duke** "History of the Wilson Football in the NFL," Wilson.com. Available at www.wilson.com/en-us/football/nfl/wilson-and-the-nfl/history/, accessed July 2, 2013.

7   **spheroid became even more prolate** Scott Oldham, "Bombs Away," *Popular Mechanics* 178, no. 10 (2001): 64–67.

8   **mind-boggling 167.01 mph** "The Formidable Record of Fred Rompelberg and Its Development." Available at www.fredrompelberg.com/en/html/algemeen/fredrompelberg/record.asp, accessed July 2, 2013.

10 **ESPN Sport Science segment** "On Sport Science, Drew Brees," ESPN Sport Science video, December 2, 2009. Available at http://youtube/watch?v=vVoqA-LKGb4, accessed July 2, 2013.

10 **using a ball rigged** John Brenkus, "Spinsanity: Revealing the Amazing Powers of a Ball in Flight," *ESPN The Magazine,* April 15, 2010.

11 **we asked William Rae** William Rae, phone interview by the authors, November 14, 2012.

11 **Myth of the Tight Spiral** William J. Rae, "Flight Dynamics of an American Football in a Forward Pass," *Sports Engineering* 6, no. 3 (2003): 149–163.

11 **do a belly flop** W. J. Rae and R. J. Streit, "Wind-Tunnel Measurements of the Aerodynamic Loads on an American Football," *Sports Engineering* 5, no. 3 (2002): 165–172.

11 **as a physicist calls it** Timothy Gay, *The Physics of Football: Discover the Science of Bone-Crunching Hits, Soaring Field Goals, and Awe-Inspiring Passes* (New York: HarperCollins, 2005).

11 **motion a complex dance** Jani Pallis, phone interview by the authors, November 13, 2012.

13 **"first-person shooter [FPS] games are sexy"** Timothy Cowan, phone interview by the authors, September 21, 2012.

13 **"a different spin"** Jerry Rice, phone interview by the authors, April 16, 2013.

14 **the ball might careen** Rod Cross, "Bounce of an Oval Shaped Football," *Sports Technology* 3, no. 3 (2010): 168–180.

15 **In a word, randomness** A. Douglas Stone, phone interview by the authors, December 19, 2012.

CHAPTER 2: TEDDY ROOSEVELT IN THE UNCANNY VALLEY

17 **Football was in trouble** J. J. Miller, *The Big Scrum: How Teddy Roosevelt Saved Football* (New York: HarperCollins, 2011).

18 **Abner Doubleday inventing baseball** Bill Deane, *Baseball Myths: Debating, Debunking, and Disproving Tales from the Diamond* (New York: Rowman and Littlefield, 2012).

18   **Yale abandoned its game** Tim Cohane, *The Yale Football Story* (New York: Putnam, 1951).

19   **rules were agreed upon** David M. Nelson, *The Anatomy of a Game: Football, the Rules, and the Men Who Made the Game* (Newark: University of Delaware Press, 1994).

20   **"Flying Wedge was hailed"** Scott A. McQuilkin and Ronald A. Smith, "The Rise and Fall of the Flying Wedge: Football's Most Controversial Play," *Journal of Sport History* 20, no. 1 (1993): 57–64.

22   **"a grand play"** John Sayle Watterson, *College Football: History, Spectacle, Controversy* (Baltimore: Johns Hopkins University Press, 2002).

22   **The game's best thinkers** A. A. Stagg and W. W. Stout, *Touchdown! As Told by Coach Amos Alonzo Stagg to Wesley Winans Stout* (New York: Longmans, Green, and Co., 1927).

25   **animating the blockbuster film *Shrek*** Lawrence Weschler, "Why Is This Man Smiling?" *Wired* 10, no. 6 (June 2002): 120. Available at www.wired.com/wired/archive/10.06/face_pr.html, accessed July 2, 2013.

25   **stumbled into the Uncanny Valley** Jamie York, "Hollywood Eyes Uncanny Valley in Animation," *All Things Considered,* National Public Radio, March 5, 2010.

25   **Masahiro Mori, a Japanese roboticist** Masahiro Mori, "The Uncanny Valley," *Energy* 7, no. 4 (1970): 33–35 (in Japanese).

26   **Mori graphed the Uncanny Valley** Masahiro Mori, "The Uncanny Valley," trans. K. F. MacDorman and N. Kageki, *IEEE Robotics and Automation Magazine* 19, no. 2 (2012): 98–100.

26   **feature earned only $665,426** *Shrek* and *Polar Express,* Internet Movie Database. Available at www.imdb.com/title/tt0126029/ and www.imdb.com/title/tt0338348/, accessed July 2, 2013.

28   **Lawrence Taylor broke Joe Theismann's leg** Richard Sandomir, "20 Years Later, Theismann Revisits Replay," *New York Times,* December 26, 2005.

28   **"immoral to be"** Will Leitch, "Is Football Wrong?" *New York,* August 10, 2012.

29   **he called together representatives** Miller, *Big Scrum.*

CHAPTER 3: THE ROBUST AND FRAGILE FACE MASK OF OTTO GRAHAM

31   **quarterbacked the Cleveland Browns** Andy Piascik, *The Best Show in Football: The 1946–1955 Cleveland Browns—Pro Football's Greatest Dynasty* (Lanham, MD: Taylor Trade Publishing, 2006).

31   **called Graham the best ever** Peter King, *Sports Illustrated: Greatest Quarterbacks* (New York: Time Inc., 2000).

32   **first helmet, made of leather** "History of the Football Helmet," Past Time Sports website. Available at www.pasttimesports.biz/history .html, accessed July 5, 2013.

33   **Browns coach Paul Brown** George Cantor, *Paul Brown: The Man Who Invented Modern Football* (Chicago: Triumph Books, 2008).

35   **"a very good case study"** John Doyle, phone interview by the authors, November 19, 2012.

36   **a very complex one** Marie E. Csete and John C. Doyle, "Reverse Engineering of Biological Complexity," *Science* 295, no. 5560 (2002): 1664–1669.

36   **Two key terms in Doyle's** John C. Doyle et al., "The 'Robust yet Fragile' Nature of the Internet," *Proceedings of the National Academy of Sciences* 102, no. 41 (2005): 14497–14502.

38   **What exactly is a concussion?** Robert Cantu, phone interview by the authors, March 18, 2013.

38   **biomechanical forces include** P. McCrory et al., "Consensus Statement on Concussion in Sport—the 3rd International Conference on Concussion in Sport, Held in Zurich, November 2008," *Journal of Clinical Neuroscience* 16 (2009): 755–763.

38   **cause the brain to crash** Centers for Disease Control and Prevention, "Concussion and Mild TBI." Available from www.cdc.gov/ concussion/index.html, accessed July 5, 2013.

38   **headache, nausea, sensitivity to light** B. D. Jordan, "The Clinical Spectrum of Sport-Related Traumatic Brain Injury," *Nature Reviews Neurology* 9 (2013): 222–230.

38   **Some minor concussions resolve** Jordan, "Clinical Spectrum."

38   **the word *concussion*** P. R. McCrory, "Concussion: The History of Clinical and Pathophysiological Concepts and Misconceptions," *Neurology* 57, no. 12 (2001): 2283–2289.

39   **"cuckoo," "goofy," "slug nutty"** H. S. Martland, "Punch Drunk," *Journal of the American Medical Association* 91, no. 15 (1928): 1103–1107.

CHAPTER 4: VINCE LOMBARDI'S BEAUTIFUL MIND

40   **greatest NFL coach of all time** David Maraniss, *When Pride Still Mattered: A Life of Vince Lombardi* (New York: Simon and Schuster, 1999).

40   **He was a geek** David Maraniss, phone interview by the authors, January 4, 2012.

41 "worked our butts off" Jeré Longman, "Eagles' 1960 Victory Was an N.F.L. Turning Point," *New York Times*, January 6, 2011.

42 "world should be knowable" David Kaiser, phone interview by the authors, October 23, 2012.

43 He taught math David Maraniss interview.

44 "*scientific* way to do everything" David Maraniss interview.

45 "quarterback days are over" Mike Vandermause, "Green Bay Packers Legend Paul Hornung Credits Vince Lombardi with Salvaging Career," *Green Bay Gazette*, May 15, 2010.

47 "bread-and-butter play" Vince Lombardi, *Vince Lombardi's the Science and Art of Football*, Front Row Video, Edison, New Jersey, 1988.

50 understood through game theory Roberto Serrano, phone interview by the authors, January 14, 2012.

50 "a lot like matching pennies" Matthew Jackson, phone interview by the authors, January 14 and April 9, 2012.

51 Lombardi's payoff matrix Avinash K. Dixit and Barry J. Nalebuff, *Thinking Strategically: The Competitive Edge in Business, Politics, and Everyday Life* (New York: W. W. Norton, 1993).

52 the maximin strategy Ken Binmore, *Game Theory: A Very Short Introduction* (Oxford, U.K.: Oxford University Press, 2007).

52 called a Nash Equilibrium John Nash, "*Non-Cooperative Games,*" *Annals of Mathematics*, Second Series, vol. 54, no. 2 (September 1951): 286–295.

53 "Machiavelli was a game theorist" John Nash, email interview by Ainissa Ramirez, February 20, 2013.

CHAPTER 5: DARWIN'S PLACEKICKER: SURVIVAL OF THE FLATTEST

54 an unspeakable accident Steve Springer, "Agajanian Made It by Half a Foot: Kicker Beat Bad Injury to Work 40 Years in NFL," *Los Angeles Times*, July 11, 1986.

55 boot was more than just right Wayne Coffey, "Friends Lobbying for Ben Agajanian, Oldest Living NY Giant, to Earn Spot in Pro Football Hall of Fame," *Daily News*, January 21, 2012.

55 Agajanian make history Bill Dwyre, "Kicking Pioneer Makes His Case," *Los Angeles Times*, October 27, 2007.

55 football's ultimate outsiders Stefan Fatsis, *A Few Seconds of Panic: A 5-Foot-8, 170-Pound, 43-Year-Old Sportswriter Plays in the NFL* (New York: Penguin, 2008).

58 "meat of your foot" Jim Breech, phone interview by the authors, March 11, 2013.

59 **grew up in eastern Poland** Fred Bednarski, phone interview by the authors, March 3, 2013.

61 **the name Ivan Putski** Thomas Jones, "American Ideal," *Westlake Picayune,* July 3, 2003.

61 **too far ahead of his time** Chip Stewart, "The Big Kick," *Horns Illustrated,* 2004, 40–45.

61 **appropriate credit for his historic kick** Dan Daly, "Going Sideways into History; Bednarski's Soccer-Style Kick in '57 Heralded a New Era," *Washington Times,* October 17, 2007.

61 **most contemporary news stories** Kevin Sherrington, "Often Overlooked, Texas' Bednarski Is the True Pioneer of Soccer-Style Kick," *Dallas Morning News,* December 8, 2012.

63 **"feels so easy"** Jim Breech interview.

64 **his toe, about 40 mph** Ryan D. Hartschuh, "Physics of Punting a Football," Physics Department, College of Wooster, Wooster, Ohio, May 2002.

64 **attained a speed of 70 mph** Gay, *Physics of Football.*

66 **"show it in stressful situations"** Sian Beilock, phone interview by the authors, April 8, 2013.

68 **performs another overarching function** Sian Beilock, *Choke: What the Secrets of the Brain Reveal About Getting It Right When You Have To* (New York: Free Press, 2010).

CHAPTER 6: THE BUTTERFLY EFFECT OF GREG COOK

74 **"John Elway before John Elway"** Geoff Hobson, "Mates Remember Cook," Cincinnati Bengals website, January 27, 2012. Available at www.bengals.com/news/article-1/Mates-remember-Cook/a3ef0a6b-3add-4923-a1a4-bbcccb522b57, accessed July 5, 2013.

74 **"NFL rookie like him"** Paul Zimmerman, "Landmark Meeting," *Sports Illustrated,* October 10, 2001.

76 **a meteorologist at MIT** James Gleick, *Chaos: Making a New Science* (New York: Viking, 1987).

76 **a tiny change in his own input** Stephen Wolfram interview.

76 **0.000127 had changed everything** Ziauddin Sardar and Iwona Abrams, *Introducing Chaos: A Graphic Guide* (London: Icon Books, 2008).

77 **called it the Butterfly Effect** Stephen Wolfram, *A New Kind of Science* (Champaign, IL: Wolfram Media, 2002), 304–315.

77 **looked like, well, a math professor** Virgil Carter, phone interview by the authors, December 21, 2012.

CHAPTER 7: AS MEL BLOUNT CHANNELS THOMAS EDISON

81   **I never had a chance** Darryl Stingley, "Darryl Stingley: Happy to Be Alive," *Ebony* 38, no. 12 (1983): 68–74.

81   **He remained a quadriplegic** Darryl Stingley with Mark Mulvoy, *Darryl Stingley: Happy to Be Alive* (New York: Beaufort Books, 1983).

82   **Chapman collapsed** "Beaned by a Pitch, Ray Chapman Dies," *New York Times,* August 17, 1920.

83   **transformation that an animal species undergoes** Charles Darwin, *On the Origin of Species* (London: John Murray, 1859).

83   **"see it in our lifetimes"** Jonathan Weiner, phone interview by the authors, April 1, 2012.

84   **evolutionary pendulum on the Galapagos** Jonathan Weiner, *The Beak of the Finch: A Story of Evolution in Our Time* (New York: Vintage, 1995).

85   **"as safe as we could make it"** Don Shula, phone interview by the authors, April 4, 2013.

86   **"the first big cornerbacks"** Ken Anderson, phone interview by Allen St. John, February 28, 2013.

88   **colorfully, the Muckers** G. S. Bryan, *Edison, the Man and His Work* (Garden City, NY: Garden City Publishing Co., 1926).

90   **"still our number one concern"** Don Shula interview.

CHAPTER 8: HOW IS A QUARTERBACK LIKE YOUR LAPTOP?

92   **Schrödinger's Cat is a thought experiment** E. Schrödinger, "Die gegenwärtige Situation in der Quantenmechanik," *Die Naturwissenschaften* 23, no. 48 (1935).

92   **What is the state of the cat?** John D. Trimmer, "The Present Situation in Quantum Mechanics: A Translation of Schrödinger's 'Cat Paradox' Paper," *Proceedings of the American Philosophical Society* 124, no. 5 (1980): 323–338.

94   **Ken Anderson was a math major** Ken Anderson interview.

94   **"five receivers out in a pattern"** Virgil Carter interview.

95   **"1–2–3 Guarantee"** Sam Wyche interview.

97   **called Boolean algebra** George Boole, *An Investigation of the Laws of Thought,* 1854.

97   **clever graduate student from MIT** Claude E. Shannon, "A Symbolic Analysis of Relay and Switching Circuits," *Electrical Engineering* 57, no. 12 (1938): 713–723.

97   **circuits that Bell Labs used** Claude Elwood Shannon and Warren

Weaver, "A Mathematical Theory of Communication," American Telephone and Telegraph Co., 1948.

99   **"sprint sixty to eighty yards"** Jerry Rice interview.

100   **forgotten coach named Sid Gillman** Jeremy Stoltz, "Chalk Talk: West Coast Offense Part I," Bear Report website, April 25, 2007. Available at http://chi.scout.com/2/638740.html, accessed July 5, 2013.

101   **illustrates his influence** Jonathan Rand, *The Year That Changed the Game: The Memorable Months That Shaped Pro Football* (Dulles, VA: Potomac Books Inc., 2008).

101   **pass patterns could be best** Tom Bass, phone interview by the authors, February 15, 2013.

102   **"based on precision"** Ken Anderson interview.

103   **"like a dance"** Jerry Rice interview.

104   **"wasn't random chance"** Virgil Carter interview.

CHAPTER 9: SAM WYCHE AT PLAY IN THE FIELDS OF CHAOS

105   **"Jeepers, Skeets"** Sam Wyche interview.

106   **Gallaudet University teammates** Jack R. Gannon, Jane Butler, and Laura-Jean Gilbert, *Deaf Heritage: A Narrative History of Deaf America* (Silver Spring, MD: National Association of the Deaf, 1981).

107   **origin of the huddle** LA84 Foundation, "The Huddle Debate Continues," *College Football Historical Society Newsletter*, February 1998, 4–5.

108   **called a "nickel period"** Sam Wyche interview.

110   **And his thumbs?** Sam Wyche interview.

110   **In a word, chaos** Wolfram, *New Kind of Science*, 304–315.

110   **creator of the Butterfly Effect** Gleick, *Chaos*.

111   **"unattainable clarity and insight"** Steven Levy, "The Man Who Cracked the Code to Everything . . . ," *Wired* 10, no. 6 (June 2002).

111   **"knob of chaos theory"** Stephen Wolfram interview.

113   **"before the clock runs down"** Gerald Eskenazi, "Bengals Accused of Pulling a Fast One: No-Huddle, Hurry-Up Offense Angers Opponents," *New York Times,* January 7, 1989.

114   **"kind of the perfect storm"** Sam Wyche interview.

114   **profound reimagining of the game** John Breech, "How the Cincinnati Bengals Changed NFL History: The No-Huddle Offense Turns 25," Bleacher Report website. Available at http://bleacherreport

.com/articles/201148-how-the-cincinnati-bengals-changed-nfl
-history-the-no-huddle-offense, accessed July 5, 2013.

CHAPTER 10: PLAYING DEFENSE, HEISENBERG STYLE

119   **passing offense is the zone blitz** John Breech, "How the Cincinnati
       Bengals Changed NFL History Part II: The Zone Blitz," Bleacher
       Report website. Available at https://bleacherreport.com/
       articles/214337-how-the-cincinnati-bengals-changed-nfl-
       history-part-ii-the-zone-blitz, accessed July 5, 2013.
119   **"football bedlam"** Chris Brown, "Controlled Chaos," *Grantland*,
       September 26, 2012. Available at www.grantland.com/story/_/id/
       8428129/dick-lebeau-evolution-coverage-tactics-zone-blitz,
       accessed July 5, 2013.
120   **"Fixin' to Score"** Brown, "Controlled Chaos."
120   **two vastly different approaches** Tim Layden, *Sports Illustrated
       Blood, Sweat, and Chalk: The Ultimate Football Playbook: How
       the Great Coaches Built Today's Game* (New York: Time Home En-
       tertainment, 2010).
120   **"defend the whole field"** Ron Jaworski, David Plaut, Greg Cosell,
       and Steve Sabol, *The Games That Changed the Game: The Evolu-
       tion of the NFL in Seven Sundays* (New York: Random House,
       2010).
125   **New York's Devin Thomas** Ann Killion, "Two Decades After
       Craig's Fumble, 49ers' Williams Joins Lonely Lore," *Sports Illus-
       trated*, January 23, 2012.
126   **swooped in and grabbed** "The Uncertainty Principle," Stanford En-
       cyclopedia of Philosophy, October 8, 2001. Available at www
       .science.uva.nl/~seop/entries/qt-uncertainty/, accessed July 5, 2013.
126   **"We were just like"** Benjamin Wallace-Wells, "Did Giants Strategi-
       cally Concuss Kyle Williams?" *New York*, January 23, 2012.
126   **"'put a hit on that guy'"** Steve Politi, "Giants vs. 49ers: Jacquian
       Williams, Devin Thomas Become Unlikely Heroes for Giants," *New
       Jersey Star Ledger*, January 23, 2012.
127   **"put a lick on him"** Sean Pamphilon, *The United States of Football*,
       docmentary feature, 120 minutes, 2012.
128   **Bountygate, do the math** Peter King, "Way Out of Bounds," *Sports
       Illustrated*, March 12, 2012.
129   **his defensive players to riff** Kevin Van Valkenburg, "Power Mad,"
       ESPN The Magazine, December 28, 2012.

CHAPTER 11: HOW TO TURN A BIG MAC INTO AN OUTSIDE LINEBACKER

133   **has the distinction** David S. Neft, Richard M. Cohen, and Rick Korch, *The Football Encyclopedia: The Complete History of Professional Football from 1892 to the Present* (New York: St. Martin's Press, 1994).

134   **By 1990 there were ninety-four** Eddie Pellis, "Number of 300-Pound NFL Linemen Ballooning," Associated Press, August 8, 2010.

134   **"America's mascot"** Tom Friend, "How 'The Fridge' Lost His Way," *ESPN.com*, February 8, 2011. Available at http://sports.espn.go.com/ nfl/playoffs/2010/news/story?id=6091766, accessed July 5, 2013.

134   **middle Paleolithic period** Richard Steckel, phone interview by the authors, November 6, 2012.

135   **using fire to cook food** Robert Wrangham, *Catching Fire: How Cooking Made Us Human* (New York: Basic Books, 2009).

135   **at the foot of geometry** Galileo Galilei, *Dialogues Concerning Two New Sciences*, 1638.

135   **mass will increase significantly** Lois H. Gresh and Robert Weinberg, *The Science of Superheroes* (Hoboken, NJ: Wiley, 2003).

135   **mass will increase significantly** T. McGraw, T. Kawai, and J. Richards, "Allometric Scaling for Character Design," *Computer Graphics Forum* 30, no. 1 (2011): 153–168.

135   **"surface in proportion to volume"** J. B. S. Haldane, *On Being the Right Size*, 1926.

136   **"steps and try to block people"** Sam Wyche interview.

137   **"working in a phone booth"** Sam Wyche interview.

138   **"human desires is alchemy"** David Katz, phone interview by the authors, March 5 and April 2, 2013.

141   **"God doesn't readily change"** Charles Yesalis, phone interview by the authors, March 3, 2013.

142   **"agreed with him twenty years ago"** Jay Hoffman, phone interview by the authors, March 15, 2013.

142   **interviewed elite strength athletes** Charles E. Yesalis and Virginia S. Cowart, *The Steroids Game: An Expert's Inside Look at Anabolic Steroid Use in Sports* (Champaign, IL: Human Kinetics Publishers, 1998).

142   **more comprehensive strength training** Jay R. Hoffman et al., "Position Stand on Androgen and Human Growth Hormone Use," *Journal of Strength and Conditioning Research* 23, no. 5 Supplement (2009): S1–S59.

CHAPTER 12: SIR ISAAC NEWTON'S FANTASY FOOTBALL DRAFT

146   "quantity of motion" Gay, *Physics of Football.*

147   How long is a hundredth Allen St. John, "Economics of the NFL Draft: How 1/100th of a Second Made Chris Johnson $53 Million," Forbes.com, April 24, 2012. Available at www.forbes.com/sites/allenstjohn/2012/04/24/economics-of-the-nfl-draft-how-1100th-of-a-second-made-chris-johnson-53-million/, accessed July 5, 2013.

151   the Olympics would be his last Frank Luksa, "Hayes Deserves Better Place in History," ESPN.com, March 28, 2008. Available at http://sports.espn.go.com/nfl/columns/story?columnist=luksa_frank&id=3312940, accessed July 5, 2013.

153   weighed in at *386 pounds* "1999–2013 Combine Results," NFL Combine Results website. Available at http://nflcombineresults.com/nflcombinedata.php, accessed July 5, 2013.

155   giant left tackles assigned Michael Lewis, *The Blind Side* (New York: W. W. Norton, 2009).

CHAPTER 13: CHOOSING YOUR NEXT QUARTERBACK?
THEY HAVE AN APP FOR THAT

157   "part cerebral talent" Paraag Marathe, phone interview by the authors, April 3, 2013.

158   "Clarity of mind" Charlie Ward, phone interview by the authors, February 22, 2013.

158   "quarterback like Ward has to" Charles Greer, phone interview by the authors, January 4 and April 9, 2013.

160   What would four pads cost? "So, How Do You Score?" ESPN.com. Available at http://espn.go.com/page2/s/closer/020228test.html, accessed July 5, 2013.

161   "make sure Grandma could cross" Jocelyn Faubert, phone interview by the authors, February 19, 2013.

163   The task of finding Neurotracker test taken by Allen St. John, Boston, Massachusetts, March 2, 2013.

CHAPTER 14: OF RISK, INNOVATION, AND COACHES
WHO BEHAVE LIKE MONKEYS

167   lost their first two games Amy Shipley, "Wildcat Package Brings Direct Results in Miami," *Washington Post*, October 18, 2008.

167   up to the front of the plane Toni Monkovic, "The Dolphins' Formation: How They Did It," *New York Times*, September 22, 2008.

168   **offense at the University of Arkansas** Steven Wine, "Dolphins Help Single Wing Make Comeback," *USA Today,* October 9, 2008.

169   **Werner Heisenberg would understand** Werner Heisenberg, *Physics and Philosophy: The Revolution in Modern Science* (New York: Harper and Row, 1958).

170   **"that didn't quite satisfy me"** Brian Burke, phone interview by the authors, March 13, 2013.

171   **Prospect theory argues that** Daniel Kahneman and Amos Tversky, "Prospect Theory: An Analysis of Decision Under Risk," *Econometrica* 47, no. 2 (1979): 263–291.

171   **found a twenty-dollar parking ticket** Keith Chen, phone interview by the authors, April 5, 2013.

172   **were throwing away strokes** Devin G. Pope and Maurice E. Schweitzer, "Is Tiger Woods Loss Averse? Persistent Bias in the Face of Experience, Competition, and High Stakes," *American Economic Review* 101, no. 1 (2011): 129–157.

173   **Wildcat offense confused the Patriots** Monkovic, "Dolphins' Formation."

174   **taught monkeys to use money** Venkat Lakshminaryanan, M. Keith Chen, and Laurie R. Santos, "Endowment Effect in Capuchin Monkeys," *Philosophical Transactions of the Royal Society B: Biological Sciences* 363, no. 1511 (December 12, 2008): 3837–3844.

175   **risk to the Monkey Market** Allen St. John, "What Monkeys Can Teach You about Money," *Mental Floss,* September–October 2011.

175   **to avoid a loss than to gain** Steven D. Levitt and John A. List, "Homo Economicus Evolves," *Science* 319, no. 5865 (February 15, 2008): 909–910.

CHAPTER 15: DESPERATION PLUS INSPIRATION EQUALS 16,632 ELIGIBLE RECEIVERS

177   **he saw on the whiteboard** Kurt Bryan, phone interview by the authors, January 23, 2013.

177   **devoted San Francisco 49ers fan** Steven Humphries, phone interview by the authors, January 23, 2013.

183   **16,632 possibilities** Here is the math behind the A-11, provided by John Matson, associate editor at *Scientific American* (used with permission):

In an ordinary offense you'd have 11 players total, of whom 5 are linemen (and therefore can't receive the ball). But any one of the remaining 6 players can receive the snap, and can then keep it themselves (a quarterback sneak, essentially) or can pass/hand off to a teammate. So there are 36 possible outcomes: 6 players who can take the snap times 6 players who can end up with it.

Here's where the magic of A-11 comes into play. The two sets of players (linemen vs. non-linemen) aren't fixed—for any given play, any one of the 11 players can be a lineman or can play one of the other positions on the field. So before you even get to those 36 options mentioned above, you have a huge number of options in deciding who's a lineman and who's not—we'll call the 6 *non*-linemen the skill players (no offense, linemen!).

Let's say that you choose your 6 skill players one by one from your 11 players on the field. You have 11 choices for skill player #1, 10 choices for skill player #2 (since one player has already been picked), 9 choices for player #3, 8 choices for player #4, 7 choices for player #5, and 6 choices for player #6.

That's 332,640 possible permutations ($11 \times 10 \times 9 \times 8 \times 7 \times 6$). But again, with the A-11 offense any one of those skill players can take the snap, and any one of them can end up with the ball. So we don't really care who is skill player #1 and who is skill player #4—they're all interchangeable. In other words, as far as who will end up with the ball, these two sets of skill players are identical: [Alex, Bob, Carl, Dave, Ed, Frank] and [Dave, Carl, Frank, Bob, Alex, Ed]. The order doesn't matter. So you have to divide the huge number of 332,640 possible permutations for the skill players by the number of times each group of players repeats as an identical group with different ordering. For each grouping of 6 skill players, they can be ordered $6! = (6 \times 5 \times 4 \times 3 \times 2 \times 1) = 720$ ways, since within each group there are 6 ways to pick the first skill player, 5 ways to pick the second skill player, and so on.

So divide those 332,640 possible permutations by the 720 ways that each unique grouping of 6 skill players (say, Alex, Bob, Carl, Dave, Ed and Frank) gets repeated: $332,640 \div 720 = 462$.

All of which is to say, with 11 players on the field, there are 462 unique ways to break the team up into 5 linemen and 6 skill players. In combinatorial jargon you'd say, "11 choose 6" = 462.

Then, for each of those 462 possible setups of offensive players, you once again have 36 possible outcomes (as in a standard offense) as far as who takes the snap and who ends up with the ball. That's where the number cited in the article comes from: $462 \times 36 = 16,632$.

184   **attracted national attention** Richard Morgan, "California High School's Offensive Scheme Adds Randomness to Football," *Scientific American*, September 2, 2008. Available at www

.scientificamerican.com/article.cfm?id=football-offensive-math, accessed July 5, 2013.

184   **attracted national attention** Josh Levin, "Could This Offense Revolutionize Football?" *New York Times,* November 1, 2008.

184   **attracted national attention** David Fleming, "The Ball Stops Here," *ESPN.com,* December 18, 2008. Available at http://sports .espn.go.com/espnmag/story?section=magazine&id=3778501, accessed July 5, 2013.

CHAPTER 16: THE MAN WHO LOVED TACKLING

188   **"Wrap him up"** Bobby Hosea, phone interview by Allen St. John, February 6, 2013.

192   **"if I had a son"** Cindy Boren, "Obama Uncertain If He'd Let a Son Play Football," *Washington Post,* January 28, 2013.

192   **Hall of Famer Harry Carson** Michael David Smith, "Harry Carson: Knowing What I Know Now, I Wouldn't Play Football," NBC Sports website, October 8, 2012. Available at http://profootballtalk .nbcsports.com/2012/10/08/harry-carson-knowing-what-i-know -now-i-wouldnt-play-football/, accessed July 5, 2013.

192   **"keep them out of youth football"** Robert Cantu interview.

193   **first class of Pee Wee players** Alan Schwarz, "Teaching Young Players a Safer Way to Tackle," *New York Times,* December 25, 2010.

195   **the story of Al Lucas** Mark Maske, "Arena Player's Death Followed a Seemingly Normal Collision," *Washington Post,* April 12, 2005.

195   **"but he was so gentle"** Andrew Dalton, "Arena Football Player Dies of Presumed Spinal Cord Injury," *San Diego Union-Tribune,* April 11, 2005.

196   **designer suits and Stacy Adams shoes** "Remembering Big Luke," Associated Press, August 28, 2005.

197   **"Al Lucas was"** Bobby Hosea interview.

CHAPTER 17: WHY WOODPECKERS DON'T GET CONCUSSIONS

199   **12,000 pecks per day** Lizhen Wang et al., "Why Do Woodpeckers Resist Head Impact Injury: A Biomechanical Investigation," *PLOS One* 6, no. 10 (2011).

199   **"small brain makes a big difference"** Lorna Gibson, phone interview by the authors, December 13, 2012.

200   **more like 100 milliseconds** Gay, *Physics of Football.*

201   **"killed three of your players"** William Simpson, phone interview by the authors, February 13, 2013.

203   "intellectual life of the whole of Western" C. P. Snow, *The Two Cultures* (Cambridge: Cambridge University Press, 1993).

203   "engineers create new things" Chuck Vest, phone interview by the authors, March 20, 2013.

203   "Look at *The Big Bang Theory*" Norman Fortenberry, phone interview by the authors, February 21, 2013.

204   "metabolic crisis going on" Robert Cantu interview.

204   "common lipid transport gene" Robert Stern, phone interview by the authors, January 1, 2013.

204   "Does a hit of 100 G equal" E. Deibert and R. Kryscio, "How Many HITS Are Too Many? The Use of Accelerometers to Study Sports-Related Concussion," *Neurology* 78, no. 22 (2012): 1712–1713.

205   CTE was found in thirty-four Robert Cantu interview.

205   This kind of basic research B. D. Jordan, "Clinical Spectrum."

205   presence of tau protein K. Kutner et al., "Lower Cognitive Performance of Older Football Players Possessing Apolipoprotein E [epsilon] 4," *Neurosurgery* 47, no. 3 (2000): 651.

206   other with tremendous force Andrew Kitchener, "An Analysis of the Forces of Fighting of the Blackbuck (*Antilope cervicapra*) and the Bighorn Sheep (*Ovis canadensis*) and the Mechanical Design of the Horn of Bovids," *Journal of Zoology* 214, no. 1 (1988): 1–20.

207   made of a porous bone Parimal Maity and Srinivasan Arjun Tekalur, "Biomechanical Analysis of Ramming Behavior in Ovis Canadensis," In *Time Dependent Constitutive Behavior and Fracture/Failure Processes, Volume 3,* ed. Tom Proulx (New York: Springer, 2011), 357–364.

207   "keratin is it deforms easily" Andrew Farke, phone interview by Ainissa Ramirez, March 27, 2013.

207   horns also bend backward P. Maity and S. A. Tekalur, "Finite Element Analysis of Ramming in Ovis Canadensis," *Journal of Biomechanical Engineering* 133, no. 2 (February 2011): 021009.

207   crumple zones in a car Andrew A. Farke, "Frontal Sinuses and Head-Butting in Goats: A Finite Element Analysis," *Journal of Experimental Biology* 211, no. 19 (2008): 3085–3094.

EPILOGUE: SHOULD THE NFL BAN HELMETS?

211   IIHS test told a different story Insurance Institute for Highway Safety, "2009 Chevrolet Malibu and 1959 Chevrolet Bel Air." Available at www.iihs.org/video.aspx/info/50thcrash, accessed July 5, 2013.

212 **"They build them better"** Insurance Institute for Highway Safety, "2009 Chevrolet Malibu and 1959 Chevrolet Bel Air."

212 **apply its methodology** Adam Bartsch, phone interview by the authors, December 12, 2012.

212 **helmet was surprisingly *bad*** A. Bartsch, E. Benzel, V. Miele, and V. Prakash, "Impact Test Comparisons of 20th and 21st Century American Football Helmets," *Journal of Neurosurgery* 116, no. 1 (2012): 222–233.

212 **high school rugby players** S. W. Marshall and R. J. Spencer, "Concussion in Rugby: The Hidden Epidemic," *Journal of Athletic Training* 36, no. 3 (2001): 334–338.

213 **5 and 6 per thousand player hours** Australian Football League Medical Officers' Association, "The Management of Concussions in Australian Football," 2011.

213 **a sixteen-year study** Randall Dick et al., "Descriptive Epidemiology of Collegiate Men's Football Injuries: National Collegiate Athletic Association Injury Surveillance System, 1988–1989 through 2003–2004," *Journal of Athletic Training* 42, no. 2 (2007): 221.

213 **"stick their faces in there"** Sean Leahy, "Ditka: Take Face Masks Away from NFL to Reduce Violent Hits," *USA Today*, October 23, 2008.

213 **"for fifteen years"** Jerry DiPaola, "Paterno, Ditka Offer Solutions to Dicey Hits," *Pittsburgh Tribune-Review*, October 20, 2010.

215 **a vehement response** A. Manickam and L. A. Marshman, "Subdural Hematoma," *Journal of Neurosurgery* 117, no. 1 (2012): 186, author reply 186–187.

215 **"that's not the answer"** Robert Stern interview.

216 **compared football to dog fighting** Malcolm Gladwell, "Offensive Play: How Different Are Dogfighting and Football?" *New Yorker*, October 19, 2009.

216 **called them paradigm shifts** T. S. Kuhn, *The Structure of Scientific Revolutions* (Chicago: University of Chicago Press, 1996).

ABOUT THE AUTHORS

An award-winning journalist and *New York Times* bestselling author, ALLEN ST. JOHN has written seven books, including *The Billion Dollar Game* and *Clapton's Guitar*. He was a columnist for *The Wall Street Journal* and *The Village Voice*, and has written for *The New York Times Magazine, U.S. News & World Report, Men's Journal, Maxim, Playboy, Road & Truck, Rolling Stone*, Esquire.com, theatlantic.com, Salon.com, and Forbes .com, where he is a regular contributor. His work has been featured in *The Best American Sports Writing*. A graduate of the University of Chicago, he lives in Montclair, New Jersey, with his wife, Sally St. John, his two children, Ethan and Emma, and Tessie the dog. (www.allenstjohn.com)

AINISSA G. RAMIREZ, PH.D., is dedicated to making science fun for people of all ages. She is the author of the TED Book *Save Our Science*, based on her TED talk on improving science education. Before that, she was a popular engineering professor at Yale University. She received her Ph.D. from Stanford University in materials science and engineering and holds several patents, one of which won her inclusion in MIT's coveted top 100 young innovators award. (www.ainissaramirez.com)

ABOUT THE TYPE

This book was set in Sabon, a typeface designed by the well-known German typographer Jan Tschichold (1902–74). Sabon's design is based upon the original letter forms of sixteenth-century French type designer Claude Garamond and was created specifically to be used for three sources: foundry type for hand composition, Linotype, and Monotype. Tschichold named his typeface for the famous Frankfurt typefounder Jacques Sabon (1520–80).